Advanced Manufacturing

Springer

London
Berlin
Heidelberg
New York
Barcelona
Budapest
Hong Kong
Milan
Paris
Santa Clara
Singapore
Tokyo

Duc Truong Pham and Ercan Oztemel

Intelligent Quality Systems

With 92 Figures

 Springer

Duc Truong Pham, PhD, DEng, CEng, FIEE
University of Wales Cardiff
School of Engineering, Systems Division,
P.O. Box 917, Cardiff CF2 1XH, UK

Ercan Oztemel, MSc, PhD
Department of Industrial Engineering, Faculty of Engineering
Sakarya University, Sakarya, Turkey

Series Editor

Professor Duc Truong Pham, PhD, DEng, CEng, FIEE
University of Wales Cardiff
School of Engineering, Systems Division,
P.O. Box 917, Cardiff CF2 1XH, UK

ISBN 3-540-76045-8 Springer-Verlag Berlin Heidelberg New York

British Library Cataloguing in Publication Data
Pham, D.T. (Duc Truong), 1952-
  Intelligent quality systems. - (Advanced manufacturing series)
  1.Quality control - Data processing 2.Adaptive control systems
  I.Title II.Oztemel, Ercan
  006.3'3
ISBN 3540760458

Library of Congress Cataloging-in-Publication Data
Pham, D. T., 1952-
  Intelligent quality systems / D.T. Pham and E. Oztemel.
      p.   cm. - - (Advanced manufacturing)
  Includes bibliographical references and index.
  ISBN 3-540-76045-8 (hardback : alk. paper)
  1. Quality control. 2. Intelligent control systems. I. Oztemel,
E. (Ercan), 1962- . II. Title. III. Series: Advanced
manufacturing series (Springer-Verlag)
TS156.P47 1996   96-8205
658.5'62 - - dc20   CIP

© Springer-Verlag London Limited 1996
Printed in Great Britain

The publisher makes no representation, express or implied, with regard to the accuracy of the information contained in this book and cannot accept any legal responsibility or liability for any errors or omissions that may be made.

Typesetting: Camera ready by authors
Printed and bound at the Athenæum Press Ltd, Gateshead
69/3830-543210 Printed on acid-free paper

Preface

Although the term quality does not have a precise and universally accepted definition, its meaning is generally well understood: quality is what makes the difference between success and failure in a competitive world. Given the importance of quality, there is a need for effective quality systems to ensure that the highest quality is achieved within given constraints on human, material or financial resources.

This book discusses Intelligent Quality Systems, that is quality systems employing techniques from the field of Artificial Intelligence (AI). The book focuses on two popular AI techniques, expert or knowledge-based systems and neural networks. Expert systems encapsulate human expertise for solving difficult problems. Neural networks have the ability to learn problem solving from examples. The aim of the book is to illustrate applications of these techniques to the design and operation of effective quality systems.

The book comprises 8 chapters. Chapter 1 provides an introduction to quality control and a general discussion of possible AI-based quality systems. Chapter 2 gives technical information on the key AI techniques of expert systems and neural networks. The use of these techniques, singly and in a combined hybrid form, to realise intelligent Statistical Process Control (SPC) systems for quality improvement is the subject of Chapters 3-5. Chapter 6 covers experimental design and the Taguchi method which is an effective technique for designing quality into a product or process. The application of expert systems and neural networks to facilitate experimental design is described in this chapter. Chapter 7 deals with inspection, a non-value-adding but often necessary quality control activity. The chapter provides examples of inspection and inspection-related tasks adopting expert system and neural network techniques. Chapter 8 discusses the monitoring and diagnosis of machines and processes and the implementation of intelligent condition monitoring and fault diagnosis systems based on expert systems and neural networks.

Much of the material in this book derives from the Quality work conducted in the authors' Intelligent Systems Laboratory at Cardiff over the past seven years.

The authors would like to acknowledge the financial support received for the work from: the Engineering and Physical Sciences Research Council (ACME Programme), the Department of Trade and Industry (Teaching Company Programme), the Higher Education Funding Council for Wales (Research Quality Programme) and the European Commission (BRITE-EURAM Programme). The work was performed as part of four collaborative projects with industrial companies including IBM, Seal Technology Systems (STS) Ltd. and Performance Vision Ltd. (PVL). The authors wish to thank their industrial partners, Mr. D. Bowen and Mr. M. Moulton (IBM), Mr. G. Davis, Mr. A. Fern and Mr. I. Ross (STS) and Mr. N. Jolliffe (PVL), for their technical and financial assistance.

Several present and former members of the Laboratory have participated in the projects. They include Mr. R. Alcock, Dr. E. Bayro-Corrochano, Dr. B. Cetiner, Mr. A. Hassan, Mr. N. Jennings, Mr. B. Peat and Dr. A. Wani, who are thanked for their contributions. The authors also extend their thanks to other Laboratory members who have helped at different stages of the work. They are Dr. P. Channon, Dr. S. Dimov, Dr. P. Drake, Dr. K. Hafeez, Mr. C. Ji and Mr. M. Packianather.

The authors are particularly indebted to Laboratory members, Mr. R. Alcock and Mr. A. Chan, for the sterling efforts they unselfishly devoted to the preparation of the book. Special thanks also go to Mr. A. Rowlands, of the Cardiff School of Engineering, for proof reading the complete manuscript.

The authors would like to thank Mr. N. Pinfield and Mrs. I. Mowbray of Springer-Verlag for their help with the production of the book.

Finally, they wish to acknowledge the permission of Wiley, Taylor and Francis and the Institution of Mechanical Engineers to use material previously published by them as detailed below:-

- Chapter 3: material from Pham, D.T. and Oztemel, E. (1992a) TEMPEX: An expert system for temperature control in an injection moulding process, *Quality and Reliability Engineering International*, 8, 9-15.

- Chapter 3: material from Pham, D.T. and Oztemel, E. (1992b) XPC: An on-line expert system for statistical process control, *Int. J. of Production Research*, 30(12), 2857-2873.

- Chapter 4: material from Pham, D.T. and Oztemel, E. (1994) Control chart pattern recognition using Learning Vector Quantisation networks, *Int. J. of Production Research*, 32(3), 721-729.

- Chapter 5: material from Pham, D.T. and Oztemel, E. (1995) An integrated neural network and expert system tool for statistical process control, *Proc. I Mech E, Part B: J. of Engineering Manufacture*, vol. 209, 91-97.

D.T. Pham
E. Oztemel

Contents

Chapter 1   Introduction

Today's manufacturing enterprises need to adopt modern tools of quality engineering to maintain and improve their competitiveness in the marketplace. The quest for quality may be attributed to the changing trends in the present consumer-oriented market where the demand for a product is often a direct derivative of its quality.

For most practical purposes, quality may be defined using variables such as cost, price and durability. A possible definition of quality is "the totality of features and characteristics of a product or service that bears on its ability to satisfy given needs" [Condra, 1993]. However, the meaning of quality may vary from person to person, depending on their position. For example, for the customer, a good quality product is one that meets his/her needs in terms of performance, appearance and price. For the product designer, quality is related to a product satisfying functional requirements. For the manufacturer, the definition of quality is based on conformance to specifications at a minimum cost.

Despite this variety of views, it is universally accepted that quality is a *customer-driven* attribute. A detailed treatise on the meaning of quality is not our intention here and readers are referred to the literature, for example [Garvin, 1984], for further discussion of this topic. This chapter explains the main features of a quality assurance system, outlines the problems of realising a quality assurance system, suggests a way forward that exploits advanced information processing tools such as expert systems and neural networks and gives a review of the state-of-the-art in intelligent quality control research.

1.1 Quality Assurance Systems

The purpose of a quality system is to determine a measure of how a product, process or machine meets the customer needs. The concept of quality is related to the functioning of an industrial organisation where all departments are required to work closely together to achieve and maintain the desired standards of quality (see Figure 1.1). In practice, usually the marketing department determines the customer requirements which form the basis of a product design in the product engineering department. It is essential that these requirements are defined as clearly as possible so that translation into actual design specifications is conducted smoothly.

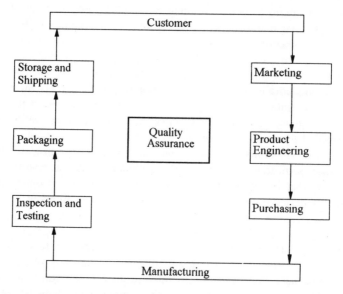

Figure 1.1 Central role of quality assurance

The role of the purchasing department is then to procure materials and components according to these specifications. Sound materials and components are the basic requirement to manufacture a quality product. The task of the manufacturing department is to set up a process which transforms the raw materials into high quality finished products. The duty of the inspection and testing department is to ensure that only good products leave the production line. The packaging department preserves and protects the product against damage during storage and shipping.

The success of a manufacturing concern depends on the deployment of a reliable quality assurance system and the establishment of effective communication links between various departments. In general, a quality assurance department is set up in each organisation whose main function is to:-

- ensure product quality and reliability;
- improve quality planning;
- control the quality levels of the suppliers;
- ensure conformance to the specifications set out;
- check and control the quality testing equipment;
- train personnel;
- provide feedback to management and the production line about quality;
- manage the in-house quality assurance system;
- evaluate the performance of the quality assurance system employed;
- maintain a keen awareness of developments in quality technology and deploy the appropriate technology to maintain and improve the competitive stance of the company.

As remarked above, the success of a quality system depends on a well-designed information flow between the various departments of an organisation. Considering the complexity, uncertainty and competitiveness of today's marketplace, the requirement for effective communication becomes even more important. The smooth functioning of advanced manufacturing technology (for example, sophisticated machine tools, manufacturing cells and systems) requires systematic methods of control, monitoring and diagnosis. Unprecedented situations and unforeseen manufacturing problems need to be resolved often using imprecise and incomplete information. The associated quality issues are to be addressed with consideration to this lack of reliable information.

A traditional way of achieving and ensuring the set quality standards is via implementing Statistical Process Control (SPC) techniques. SPC is the observation of the quality of a semi-finished or finished product by correlating it with the time histories of the manufacturing process involved and taking appropriate measures to maintain the desired quality level. However, as manufacturing complexity and uncertainty are increased, the SPC procedures become more demanding. There is a shortage of good SPC specialists as the skills to implement proper SPC procedures develop over time making use of accumulated knowledge of the process involved. Also, a specialist's skills may vary from one machine, or plant, to another and involve human factors regarding learning ability, attitude and decision making aptitude.

One way to improve SPC procedures is to replace the SPC specialist with computers which are able to mimic human-like intelligent behaviour. In recent years, attention has focussed on artificial intelligence (AI), a branch of computer science which has shown great promise in dealing with difficult manufacturing problems. What makes AI techniques popular is their ability to learn from experience and to handle uncertain, imprecise (fuzzy) and complex information in a competitive and quality demanding environment.

Among the available AI tools, knowledge-based (KB) systems or expert systems and neural networks are those which have attracted the most effort from researchers and practitioners for solving many quality assurance issues. The main architecture and decision making mechanisms of these two tools are described at length in Chapter 2. KB systems have been one of the most successful practical products of AI. The main advantage of KB systems is that they enable even a novice to solve complex problems and arrive at a conclusion nearly as good as that arrived at by a human expert. Neural networks are also gaining much attention because of their inherent abilities automatically to acquire and store knowledge and integrate information obtained from multiple sensors in real-time. In theory, the information processing time can be greatly reduced due to the parallel architecture of neural networks. This parallel architecture also gives the system the capability of being more robust, adaptive and flexible.

This book explores the use of KB systems and neural networks for manufacturing process design and control to assist quality engineers, managers and operators. This chapter presents a literature review of knowledge-based systems and neural networks as applied to quality control tasks. Since it is not our aim to write an introductory book on the subject of quality control techniques, only outline information is provided in the relevant chapters. Interested readers are referred to the literature in the field, for example [Grant and Leavenworth, 1988], for more information on such techniques.

1.2 Knowledge-Based Systems for Quality Control

Knowledge-based systems[1] have been used to solve a range of decision making problems in manufacturing and have excelled in many analysis tasks usually handled by human experts [Alexander, 1987; Yih and Nof, 1991; Pham, 1994; Leff, 1992]. Most of these systems are for planning, modelling and scheduling of manufacturing processes. Recently, the application of KB systems to quality control has become practicable due to advances in computer technology [Pham

---

[1] The terms 'knowledge-based system' and 'expert system' will be used interchangeably in this book.

and Oztemel, 1992a]. The majority of the computer programs developed for quality control are used off-line and are not intended to exercise process knowledge or provide interpretations of the process states. These functions are left to quality experts. Sanders *et al.* [1989] have highlighted the weaknesses of existing software packages indicating that their outputs can only be interpreted by very experienced operators or quality control engineers. A number of other researchers have also emphasised the need to employ knowledge-based systems as an alternative way to overcome the shortage of quality control specialists. See [Rao and Lingaraj, 1988; Iwata, 1988; Spur and Specht, 1992] for example.

There have been many attempts to use knowledge-based systems to assist quality control engineers or process operators [Ballard, 1987; Dybeck, 1987; Hubele and Keats, 1987; McFall, 1987; Gilman, 1988; Greene, 1991; Bird, 1992]. A number of authors have suggested procedures for designing knowledge-based systems and reported preliminary designs [Furukawa and Ishizu, 1985; Cai, 1986; Kamal, 1988; Hosseini and Fard, 1989; Kaya, 1989; Barsegian and Melkonian, 1988; Winchell and Chugh, 1989; Affisco and Chandra, 1990; Grob and Pfeifer, 1990; Schachter-Radig and Wermser, 1990; Crossfield and Dale, 1991; Franz and Foster, 1992].

There have been other studies addressing the technical issues involved in the development of expert systems for various quality problems including knowledge acquisition and elicitation, knowledge representation, designing explanation and user interface modules and real-time design [Bailey, 1987; Eccleson and Main, 1988; Sanders *et al.*, 1988]. For some researchers, the main aim was to show how an expert system can process uncertain quality data using fuzzy logic reasoning [Aliyev and Tserkovnyy, 1986; Chacon and Liles, 1988; Kohoutek, 1989] or qualitative reasoning [Uhrik and Mezgar, 1990; Shaw and Menon, 1990].

However, only a small number of the currently used systems are reported to be operating satisfactorily. Mostly, they are designed to maintain a set quality level rather than to improve it [Brink and Mahalingam, 1991]. Fjellheim and Opdahl [1988] have described an expert system called EPAK to maintain a stable quality for the product in a paper production plant. The system is able to evaluate the paper quality, correct deviations through adjustment of control parameters and predict the consequences of the proposed actions. Braun [1990] has reported a number of systems such as DEFT, FAITH, FIS and FALCON which have been developed for quality control purposes. DEFT is a diagnostic expert system used at IBM plants to check for machine faults and diagnose airflow problems. FAITH is a general-purpose diagnostic expert system for troubleshooting spacecraft by evaluating telemetry data transmitted from a spacecraft to the earth. FIS is a fault isolation system to assist the testing and repairing of avionics. FALCON is an expert system for chemical plants that diagnoses likely causes of quality deviations.

ASASP [Fard and Sabuncuoglu, 1990] is an advisor for the selection of attribute sampling plans for the control of part or quality characteristics. The system considers the selection of single, double or multiple sampling plans and makes recommendations responding to the need of a user. RA-IQSE [Irgens, 1990; Irgens, 1991; With and Irgens, 1990] is a feature-oriented knowledge-based system for predicting the quality of machined parts and products. The system is designed to support the design of mechanical products such that their quality is optimised through observations. Associated predictive knowledge enables access to the quality influencing factors and parameters during the design process.

Leitch and Kraft [1988] and Leitch *et al.* [1991] have developed RESCU - a real-time knowledge-based system for product quality control in a chemical plant. RESCU provides reliable estimates regarding the product quality and makes recommendations on how to improve it. It can also estimate the errors in instrumentation and supply justifications for the given recommendations.

Ogai *et al.* [1990] have described TRIOS, a quality diagnosis expert system. The system predicts the final quality properties from the operating conditions of the intermediate process and lists the appropriate measures to achieve target values. LABGEN [Klaessens *et al.*, 1989] is a quality control expert system developed to monitor the performance of a chemical laboratory. The system is able to simulate the laboratory and control its quality performance.

Owen [1988] has reported an expert system called SYNOPSIS which uses the failure mode and effect analysis (FMEA) technique. SYNOPSIS enables engineers with no experience of FMEA to use it for design reviews and quality planning. The system adopts a structured approach for reviewing designs and manufacturing methods with a view to minimising possible failures. RIC, on the other hand, is an expert system which ensures the quality of purchased materials and gives advice on material receipts and inspection strategies [Crawford and Eyada, 1989].

Browet *et al.* [1988] have developed a system to provide estimates of the quality of analytical data by monitoring both instrument performance and metal determination results produced by a spectrometer. Lee *et al.* [1989] have described a prototype expert system for designing efficient experiments. The system utilises the Taguchi experimental design methodology and is for off-line use. Chen [1989] has developed a knowledge-based system to assist meat graders. It receives the characteristics of the carcass verbally from the meat grader and determines the quality and grade of the carcass. Utilising speech processing technology for intelligent quality control has also been discussed by Cassford [1987]. Groth and Moden [1991] have developed a real-time knowledge-based system for quality control and fault diagnosis of analytical instruments in a clinical chemistry laboratory.

Inspection is an important area of quality control. The main aim of an inspection system is to detect defective parts or improper assemblies during manufacturing and help to ensure that the characteristics of the item under test conform with predefined specification standards. Inspection problems have been discussed by Jenter and Schmidberger [1982]. Usually, the inspection data is acquired using computer vision tools and several knowledge-based systems have been reported to facilitate inspection using computer vision [Solinsky, 1986; Petrovic and Hinkle, 1987; Darwish and Jain, 1991].

Expert systems have found many applications in SPC [Oztemel, 1992]. Most of the research in this area is related to the construction and analysis of control charts. For example, see [Alexander and Jagannathan, 1986; Dagli and Stacey, 1988; Hosni and Elshennawy, 1988]. There are a few authors who have concentrated on various other features of SPC involving, for example, cause and effect analysis [Bourne *et al.*, 1989] or automatic interpretation of control charts to detect out-of-control situations [Evans and Lindsay, 1988; Cesarone, 1991; Dagli and Smith, 1991].

The systems mentioned above are for providing off-line assistance to quality specialists rather than process operators as they do not incorporate process specific knowledge. They are capable of selecting suitable control charts and automatically interpreting them. User intervention is usually required in these systems to define the location of possible starting points of an out-of-control (out-of-quality) situation. Pham and Oztemel [1992b] have developed an automatic on-line SPC system capable of constructing the mean and range charts, performing capability analysis and updating these charts. An automatic pattern recognition system is also provided in order to control the long term behaviour of the process.

The topic of pattern recognition for quality control has also been addressed by several researchers. Here, expert systems are used mainly to classify various control chart patterns. Most of these systems make use of statistical hypotheses, heuristics and templates and require *a priori* process knowledge. For example, Swift [1987] has designed a knowledge-based system to recognise control chart patterns which utilises statistical hypotheses and is designed to be used off-line. One drawback of this system is that, once an out-of-quality state is detected in a process, it is not possible to bring it into the in-quality state without human intervention. Pham and Oztemel [1992c] have developed an on-line control chart pattern recogniser to solve this problem. However, the system uses assumptions and heuristics and requires process knowledge. Other systems have been reported for control chart pattern recognition using templates [Cheng, 1989; Cheng and Hubele, 1989] or control theory [Anderson *et al.*, 1989]. Some researchers have utilised control theory as opposed to control charts to overcome problems with the charts. Details of their work may be found in [Love and Simaan, 1989; Simaan and Love, 1990].

1.3 Neural Networks for Quality Control

Following a hiatus, neural networks (NNs) have recently regained popularity and have successfully been employed to solve many industrial problems [Willis *et al.*, 1991]. Several NN-based systems have been designed for quality control applications and interest in this area is growing rapidly. NNs are an attractive tool for quality control due to their ability to process large amounts of data in real-time and their capacity for handling noisy, uncertain or fuzzy process data [Venkatasubramanian and Chan, 1989]. Furthermore, NNs can learn complex non-linear and multivariate relationships between process parameters. This enables them to be used in the solution of many quality control problems. Due to their adaptability, NNs can readily optimise a process condition as well as reoptimise the same process with changed dynamics or degraded equipment. Many researchers believe that NNs can provide the intelligence needed by an advanced automated manufacturing system to improve its performance and productivity.

Neural networks have been successfully employed in various quality control tasks especially when there is a requirement to process and classify a large amount of quality control data [Smith and Dagli, 1990]. Nieuwenhuize [1991] has highlighted the trade-offs of using NNs for the quality control of a food production process. Wu *et al.* [1991] have proposed adopting neural networks for diagnosing an injection moulding process to improve its quality.

There are other examples of neural network applications for process quality control which range from the control of communication networks to the control of power systems. These systems have been developed to help quality functions such as design, diagnosis, inspection and control charting.

For example, Hiametsu [1990] has deployed a neural network for controlling the quality of communication networks where a NN was used to optimise basic call admission control routines. VerDuin [1990] has discussed the use of neural networks in a continuous process to analyse infra-red spectrometer data. Marko *et al.* [1989] have used a neural network for interpreting engine faults to replace time consuming and expensive human involvement. Their findings suggest that the developed NNs outperform their human counterparts in analysing a large amount of data with greater speed and accuracy. Similar results have been quoted by others [Watanabe *et al.,* 1989; Dietz *et al.,* 1989]. An experimental study performed by GTE for monitoring a fluorescent-bulb manufacturing process has been reported by DARPA [1988], where a NN has been used to process continuous data and perform statistical analysis. Franklin *et al.* [1989] have extended this work to real-time quality control applications in manufacturing. Sobajic *et al.* [1989] have developed an on-line monitoring and diagnostic aid for the operating conditions of a power system.

The topic of speech recognition for quality control purposes has also attracted the attention of neural network researchers. Schramm and Kolb [1990] have designed a neural network able to classify acoustic signals for inspecting the quality of glass, ceramics and electric motors. The system is reported to be adaptable and able to classify quality into 10 different grades as opposed to the two traditional classes of 'good' and 'defective'. For many products, functionality and life are a direct result of their surface condition. Several researchers have analysed the images of the surface of a product to trace defects using neural networks [Tenorio and Hughes, 1987; Glover, 1988; Pham and Bayro-Corrochano, 1994]. These images are usually obtained by employing standard computer vision tools and techniques.

There have been attempts to use neural networks for analysing SPC charts [Hruska and Kuncicky, 1991]. Pugh [1989] has applied a NN to detect shifts from normal conditions using actual process data. The NN is reported to perform as well as the control chart procedure. Guo and Dooley [1990] have presented similar results using a NN to detect the shift in the process mean or variability, quoting an accuracy of 94% in detecting out-of-control situations.

The topic of control chart pattern recognition has also attracted attention from the NN community. Hwarng and Hubele [1991] have proposed a neural-network-based recogniser for this task. The system can accept 8 process values at one time and each control chart is divided into 7 quality zones. The drawback of this system is that, as it has the facility only to process very recent measurements, it cannot memorise the longer term behaviour of the process. Pham and Oztemel [1992d] have presented a pattern recognition system based on the Multi-Layer Perceptron (MLP) network [Pham and Liu, 1995] which uses real-time continuous data from the process. The NN can take into account 60 process quality values to determine the state and behaviour of the process. Pham and Oztemel [1994] have also developed a pattern recognition system utilising Learning Vector Quantisation (LVQ) neural networks. They have found that LVQ-based pattern recognisers require less training time and effort and perform better in classification and recognition than those based on MLP networks.

1.4   Integrating Expert Systems and Neural Networks for Quality Control

Due to the architectural dissimilarities in their knowledge storage and retrieval processes and in their decision making mechanisms, expert systems and neural networks are two disparate information processing tools. On a few occasions, they have been combined to capitalise on their respective strengths [Caudill, 1991].

Calabrese *et al.* [1991] have developed ESQC, an expert system integrated with a Multi-Layer Perceptron to control the production cycle of printed circuit boards. The system utilises the MLP to carry out the pure fault diagnosis subtask and the expert system for the overall coordination task. Smith and Yazici [1992] have developed a system which uses a neural network for defining the relationships between quality variables for fault detection and an expert system to diagnose and explain why and how the process is changing its behaviour. The expert system also recommends corrective actions to bring the process to an in-control situation. Oztemel [1992] has described a composite quality control system integrating neural networks with an expert system called XPC. The NN monitors the long term behaviour of the process whereas XPC controls the short term behaviour and suggests corrective actions when the process is out-of-control. The overall control of the system is the responsibility of XPC which activates the neural network modules when required.

1.5 Summary

This chapter has provided a brief introduction to quality control and quality assurance systems. Problems in the realisation of quality assurance systems have been outlined and possible solutions have been presented. These solutions employ state-of-the-art intelligent information processing tools, namely, expert systems and neural networks. A review of the application of these technologies to quality control has been provided. The review has highlighted the fact that only a handful of expert system or neural network based quality assurance systems are in on-line use. Also, the integration of neural networks with expert systems to solve quality engineering problems has not as yet been widely studied.

References

Affisco, J.F. and Chandra, M. (1990) Quality assurance and expert systems - A framework and a conceptual model, *Expert Systems with Applications*, 1, 147-153.

Alexander, S.M. and Jagannathan, V. (1986) Advisory system for control chart selection, *Computers and Industrial Engineering*, 10(3), 171-177.

Alexander, S.M. (1987) The application of expert systems to manufacturing process control, *Computers and Industrial Engineering*, 12(4), 307-317.

Aliyev, R.A. and Tserkovnyy, A.E. (1986) An intelligent robot for quality estimation and sorting of components for automated quality control, *Soviet Journal of Comput. Syst. Sci.*, 24(3), 113-119.

Anderson, K.R., Coleman, D.E., Hill, C.R., Jaworski, A.P., Love, P.L., Spindler, D.A. and Simaan, M. (1989) Knowledge-based statistical process control, *Innovative Applications of Artificial Intelligence*, Schorr, H. and Rappaport, A. (eds.), AAAI Press, 169-180.

Bailey, M.G. (1987) RESCU: A KBS for product quality control, *Proc. Int. Conf. on Knowledge-Based Systems*, London, June 1987, 233-243.

Ballard, R. (1987) Prospects for expert systems in quality management, *CME*, January 1987, 16-18.

Barsegian, V.A. and Melkonian, A.E. (1988) One approach for designing of expert system for quality control under fuzziness of industrial information, *Cybernetics and Systems '88*, Trappl, R. (ed.), Kluwer Academic, Boston, 723-728.

Bird, S. (1992) Object-oriented expert system architectures for manufacturing quality management, *J. of Manufacturing Systems*, 11(1), 50-60.

Bourne, J., Liu, H., Orogo, C. and Uckun, S. (1989) Intelligent systems for quality control, *Proc. 3rd Int. Conf. on Expert Systems and the Leading Edge in Production and Operations Management*, Head Island, South Carolina, 21-24 May 1989, 321-332.

Braun, R.J. (1990) Turning computers into experts, *Quality Progress*, February 1990, 71-75.

Brink, J.R. and Mahalingam, S. (1991) An expert system for quality control in manufacturing, *Handbook of Expert Systems in Manufacturing*, Mouse, R. and Keyes, J. (eds), McGraw Hill, USA, 455-465.

Browet, W.R., Cox, T.D.A. and Stillman, M.J. (1988) Development of an expert system for quality control monitoring of metal analysis using automatic absorption spectroscopy, *Abstracts of Papers: American Chemical Society*, June 1988, vol. 195, 223.

Cai, Z. (1986) Some research work on expert systems in AI course at Purdue University, *IEEE Int. Conf. on Robotics and Automation*, San Francisco, California, 7-10 April 1986, vol. 3, 1980-1985.

Calabrese, C., Gnerre, E. and Fratesi, E. (1991) An expert system for quality assurance based on neural networks, *Parallel Architectures and Neural Networks, 4th Workshop, Int. Institute for Advanced Scientific Studies*, Salerno, Italy, 296-300

Cassford, G.E. (1987) Talking with computers - A quality control application, *New Frontiers in Manufacturing, Proc. 10th Annual British Robot Assoc. Conf.*, Birmingham, May 1987, 97-105.

Caudill, M. (1991) Expert networks, *Byte*, October, 108-116.

Cesarone, J. (1991) QEX: An in process quality control expert system, *Robotics and Computer Integrated Manufacturing*, 8(4), 257-264.

Chacon, G.R. and Liles, D.H. (1988) An expert fuzzy advisor for the statistical control of a manufacturing process, *MIDCON '88, Conf. on Electron Conventions Management*, Dallas, TX, August 1988, 74-77.

Chen, Y.R. (1989) Applying knowledge-based expert system to meat grading, *Artificial Intelligence Systems in Government Conf.*, Washington DC, March 1989, 120-123.

Cheng, C. (1989) *Group Technology and Expert System Concepts Applied to Statistical Process Control in Small Batch Manufacturing*, PhD dissertation. Graduate College, Arizona State University, Tempe, AZ.

Cheng, C. and Hubele, N.F. (1989) A framework for a rule-based deviation recognition system in statistical process control, *Int. Industrial Engineering Conf. IIE*, Toronto, May 1989, 677-682.

Condra, L.W. (1993) *Reliability Improvement with Design of Experiments*, Marcel Dekker, New York.

Crawford, K.A. and Eyada, O.K. (1989) A prolog-based expert system for the allocation of quality assurance program sources, *Computers and Industrial Engineering*, 17(1-4), 298-302.

Crossfield, R.T. and Dale, B.G. (1991) The use of expert systems in total quality management: An exploratory study, *Quality and Reliability Engineering International*, vol. 7, 19-26.

Dagli, C.H. and Stacey, R. (1988) A prototype expert system for selecting control charts, *Int. J. of Production Research*, 26(5), 987-996.

Dagli, C.H. and Smith, A.E. (1991) A prototype quality control expert system integrated with an optimization module, *The World Congress on Expert Systems Proc.*, Orlando, Florida, 16-19 December 1991, vol. 3, 1959-1966.

DARPA (1988) *Neural Network Study*, Fairfax, VA: AFCEA International, pp. 411-415

Darwish, A.M. and Jain, A.K. (1991) A rule-based approach for usual pattern inspection, *Computer Vision Advances and Applications*, Kontri, R. and Jain, R. C. (eds.), IEEE Computer Soc. Press, Los Alamutos, California, pp. 618-630.

Dietz, W.E., Kiech, E.L. and Ali, M. (1989) Jet and rocket engine fault diagnosis in real time, *Journal of Neural Computing*, 1(1), 5-18.

Dybeck, M. (1987) Taking process automation one step further: SPC, *6th Annual Control Engineering Conf.*, Rosemount, May 1987, 643-651.

Eccleson, P. and Main, R. (1988) A knowledge-based system for manufacturing quality control, *Research and Developments in Expert Systems V, Proc. of Expert Systems '88, 8th Annual Technical Conf. of British Computer Society*, Brighton, 12-15 December 1988, pp. 245-254.

Evans, J.R. and Lindsay, W.M. (1988) A framework for expert system development in statistical quality control, *Computers and Industrial Engineering*, 14(3), 335-343.

Fard, N.S. and Sabuncuoglu, I. (1990) An expert system for selecting attribute sampling plans, *Int. J. of Computer Integrated Manufacturing*, 3(6), 364-372.

Fjellheim, R. and Opdahl, P.O. (1988) An expert system for quality control in paper production, *Artificial Intelligence III: Methodology, Systems and Applications, 3rd Int. Conf.*, Bulgarian Academy of Science, Varna, Bulgaria. September 1988, pp. 415-423.

Franklin, J.A., Sutton, R.S., Anderson, C.W., Selfridge, O.G. and Schwartz D.B. (1989) Connectionist learning control at GTE laboratories, *SPIE - Intelligent Control and Adaptive Systems*, vol. 1196, pp. 242-253.

Franz, L.S. and Foster, S.T. (1992) Utilizing knowledge-based decision support system as total quality management, *Int. J. of Production Research*, 30(9), 2159-2171.

Furukawa, O. and Ishizu, S. (1985) An expert system for adaptive quality control, *Int. Journal of General Systems*, vol. 11, pp. 183-199.

Garvin, D.A. (1984) What does product quality mean?, *Sloan Management Review*, 26(1), 25-43.

Gilman, W.R. (1988) Redefining co-ordinate metrology, *Quality*, 27(4), 20-24.

Glover, D.D. (1988) A hybrid optical fourier/electronic neurocomputer machine vision inspection system, *ROBOTICS 12, Vision '88 Conf. Proc.*, June 5-9 1988, Detroit, MI SME, vol. 2, pp.8-77 - 8-103

Grant, E.L. and Leavenworth, R.S. (1988) *Statistical Quality Control*, McGraw Hill, New York, 6th edition.

Greene, R.T. (1991) Software architecture to support total-quality companies, *Handbook of Expert Systems in Manufacturing*, Mouse, R. and Keyes, J. (eds), McGraw Hill, New York, pp. 467-487.

Grob, R. and Pfeifer, T. (1990) Knowledge-based fault analysis as a central component of quality assurance, *Methods of Operations Research*, vol. 63, pp. 545-554.

Groth, T. and Moden, H. (1991) Knowledge-based system for real-time quality control and fault diagnosis of multitest analyzers, *Computer Methods and Programs in Biomedicine*, vol. 34, pp. 175-190.

Guo, Y. and Dooley, K.J. (1990) The application of neural networks to a diagnostic problem in quality control, *Symp. on Monitoring and Control of Manufacturing Processes*, ASME Annual Meeting, Dallas, TX, November 1990, pp. 111-122.

Hiametsu, A. (1990) ATM communication networks control by neural networks, *IEEE Trans. on Neural Networks*, 1(1), 122-130.

Hosni, Y.A. and Elshennawy, A.K. (1988) Quality control and inspection - Knowledge-based quality control system, *Computers and Industrial Engineering*, 15(1-4), 331-337.

Hosseini, J. and Fard, N.S. (1989) Conceptualization and formalization of knowledge in a knowledge-based quality assurance system, *Proc. 3rd Int. Conf. on Expert Systems and the Leading Edge in Production Management*, Head Island, South Carolina, 21-24 May 1989, pp 333-341.

Hruska, S.I. and Kuncicky, D. (1991) Application of two-stage learning to an expert network for control chart selection, *Intelligent Engineering Systems Through Artificial Neural Networks, Proc. of the Artificial Neural Networks in Engineering (ANNIE'91) Conf.*, St Louis, Missouri, 10-13 November 1991, pp. 915-920.

Hubele, N.F. and Keats, J.B. (1987) Automation - the challenge for SPC, *Quality*, 26(3), 14-22.

Hwarng, H.B. and Hubele, N.F. (1991) X-Bar chart pattern recognition using neural nets, *45th Annual Quality Congress, American Society for Quality Control*, Milwaukee, 20-22 May 1991, 884-889.

Irgens, C. (1990) RA-IQSE: A system for on-line quality support for the designer of machined parts and products, *Computer Integrated Manufacturing Systems*, 3(4), 246-251.

Irgens, C. (1991) A feature-based KBS for quality prediction of machined parts and products, *Computer Integrated Manufacturing, 7th CIM European Annual Conf.*, Turin, Italy, May 1991, pp. 385-396.

Iwata, K. (1988) Applications of expert systems to manufacturing in Japan, *Int. J. of Advanced Manufacturing Technology*, 3(3), 23-37.

Jenter, W. and Schmidberger, E.J. (1982) Solving problems of industrial quality control by means of image processing systems, *Proc. of the 2nd Int. Conf. on Robot Vision and Sensory Control*, 2-4 November 1982, Stuttgart, Germany, pp. 367-378.

Kamal, M.A. (1988) Building expert systems for statistical process control in steel remelting industry, *Focus on Software - 16th Annual Computer Science Conf. ACM*, Atlanta, GA, February 1988, pp. 197-205.

Kaya, A. (1989) An intelligent supervisory system for on-line statistical process control, *Proc. of 28th Conf. on Decision and Control*, Florida, December 1989, pp. 781-782.

Klaessens, J., Sanders, J., Vandeginste, B. and Kateman, G. (1989) LABGEN: expert system for knowledge-based modelling of analytical laboratories - part 2 - Application to a laboratory for quality control, *Analytica Chimica Acta*, vol. 222, pp. 19-34.

Kohoutek, H.J. (1989) Fuzzy logic approach to IC quality assurance plan selection, *Quality and Reliability Engineering International*, vol. 5, pp. 267-272.

Lee, N.S., Phadke, M.S. and Keny, R. (1989) An expert system for experimental design in off-line quality control, *Expert Systems*, 6(4), 238-249.

Leff, L. (1992) Artificial intelligence newsletter, *Artificial Intelligence in Engineering*, vol. 7, pp. 111-119.

Leitch, R.R. and Kraft, R. (1988) A real-time knowledge-based system for product quality, *IEEE Int. Conf. on Control*, London, 13-15 April 1988, pp. 281-286.

Leitch, R.R., Kraft, R. and Luntz, R. (1991) RESCU: A real-time knowledge-based system for process control, *IEE Proc. Part D*, 138(3), 217-227.

Love, P.L. and Simaan, M. (1989) A knowledge-based system for the detection and diagnosis of out-of-control events in manufacturing processes, *American Control Conf.*, Pittsburgh, PA, June 1989, vol. 3, pp. 2394-2399.

Marko, K.A., James, J., Dosdall, J. and Murphy, J. (1989) Automotive control system diagnostics using neural nets for rapid pattern classification of large data sets, *Proc. Int. Joint Conf. On Neural Networks*, Washington DC, 2, 18-22 June 1989, vol. II, pp. 13-16.

McFall, M. (1987) Expert systems - Computer-aided quality assurance, *Quality*, 26(9), 14-16.

Nieuwenhuize, C.M.M. (1991) The use of a neural network for quality control, *Graduation Report: Faculty of Applied Physics*, Delft University of Technology, Lorentzweg, Delft, November 1991.

Ogai, H., Ueyama, T., Sato, H., Mishima, Y. and Itonaga, S. (1990) An expert system for quality diagnosis of glass films on silicon steels, *ISIJ International*, 30(2), 173-181.

Owen, B.D. (1988) An expert system to aid quality planning, *4th Int. Conf. on Computer-Aided Production Engineering*, Edinburgh, UK, November 1988, 529-534.

Oztemel, E. (1992) *Integrating Expert Systems and Neural Networks for Intelligent On-line Statistical Process Control*, PhD thesis, University of Wales, College of Cardiff, UK.

Petrovic, D. and Hinkle, E. B. (1987) A rule-based system for verifying engineering specifications in industrial visual inspection applications, *IEEE Transactions on Pattern Analysis and Machine Intelligence*, vol. PAMI-9 (2), 306-311.

Pham, D.T. (1994) Artificial intelligence in engineering, *Proc. 1994 Conf. on Applications of Artificial Intelligence Techniques in Engineering*, invited lecture, October 1994, Naples, Italy, 5-37.

Pham, D.T. and Bayro-Corrochano, E.J. (1994) Neural classifiers for automated visual inspection, *Proc. IMechE, J. of Automobile Engineering*, vol. 208, part D, 83-89.

Pham, D.T. and Liu, X. (1995) *Neural Networks for Identification, Prediction and Control*, Springer Verlag, Berlin and London, Chapter 1.

Pham, D.T. and Oztemel, E. (1992a) TEMPEX: An expert system for temperature control in an injection moulding process, *Quality and Reliability Engineering International*, vol. 8, pp. 9-15.

Pham, D.T. and Oztemel, E. (1992b) XPC: An on-line expert system for statistical process control, *Int. J. of Production Research*, 30(12), 2857-2873.

Pham, D.T. and Oztemel, E. (1992c) A knowledge-based statistical process control system, *2nd Int. Conf. on Automation, Robotics and Computer Vision*, Singapore, 16-18 September 1992, vol. 3, pp. INV-4.2.1 - INV-4.2.6.

Pham, D.T. and Oztemel, E. (1992d) Control chart pattern recognition using neural networks, *J. of Systems Engineering*, Special Issue on Neural Networks, 2(4), pp. 256-262.

Pham, D.T. and Oztemel, E., (1994) Control chart pattern recognition using learning vector quantisation networks, *Int. J. of Production Research,* 32(3), 721-729.

Pugh, G.A. (1989) Synthetic neural networks for process control, *Computers and Industrial Engineering*, 17(1-4), 24-26.

Rao, H.R. and Lingaraj, B.P. (1988) Expert systems in production and operations management: Classification and prospects, *Interfaces*, 18(6), 80-91.

Sanders, B.E., Sanders, S.A.C., Sharma, M.R. and Cherrington, J.E. (1988) From task analysis to knowledge-based system in the diagnostics and control of the injection moulding process, *COMADEM, 1st UK Seminar on Condition Monitoring and Diagnostics Engineering Management*, Birmingham, UK, September 1988, pp. 376-380.

Sanders, B.E., Sanders, S.A.C. and Cherrington, J.E. (1989) Knowledge acquisition and knowledge-based systems as an aid to product quality control, *Advances in Manufacturing Technology IV, 5th National Conf. on Production Research*, Huddersfield, UK, September 1989, pp. 364-368.

Schachter-Radig, M.J. and Wermser, D. (1990) Quality assurance as a dynamical production process guide - control elements supported by dedicated knowledge-based systems, *Computer Integrated Manufacturing*, Floria, L. and Puymbroeck, V. (eds), Springer Verlag, UK, 15-17 November 1990, pp. 233-244.

Schramm, H. and Kolb, H.J. (1990) Acoustic quality control using a multi-layer neural net, *ISATA: 22nd Int. Symp. on Automotive Technology and Automation*, Florence, Italy, 14-18 May 1990, vol. 1, pp. 709-714.

Shaw, M.J. and Menon, U. (1990) Knowledge-based manufacturing quality management: A qualitative reasoning approach, *Decision Support Systems*, 6(1), 59-81.

Simaan, M. and Love, P.L. (1990) Knowledge-based detection of out-of-control outputs in process control, *Proc. of the 29th IEEE Conf. on Decision and Control*, Honolulu, Hawaii, December 1990, pp. 128-129.

Smith, A.E. and Dagli, C.H. (1990) *Backpropagation Neural Network Approaches to Process Control: Evaluation and Comparison*, Technical Report #90-22-47, Dept. of Engineering Management, University of Missouri-Rolla.

Smith, A.E. and Yazici, H. (1992) An intelligent composite system for statistical process control, *Engineering Applications of Artificial Intelligence*, 5(6), 519-526.

Sobajic, D.J., Pao, Y.H. and Dolce, J. (1989) On-line monitoring and diagnosis of power system operating conditions using artificial neural networks, *ISCAS '89, 22nd IEEE Int. Symp. on Circuits and Systems*, Portland, OC, vol. 3, pp. 2243-2246.

Solinsky, J.C. (1986) The use of expert systems in machine vision recognition, *Vision '86 Conf. Proc.*, 3-5 June 1986, Detroit, MI, vol. 4, pp. 139-157.

Spur, G. and Specht, D. (1992) Knowledge engineering in manufacturing, *Robotics and Computer Integrated Manufacturing*, 9(4/5), 303-309.

Swift, J.A. (1987) *Development of a Knowledge-based Expert System for Control Chart Pattern Recognition Analysis*, PhD dissertation, Graduate College, Oklahoma State University, Stillwater, Oklahoma.

Tenorio, M. F. and Hughes, C.S. (1987) Real-time noisy image segmentation using an artificial neural network model, *Proc. of the IEEE First Int. Conf. on Neural Networks*, San Diego, California, vol. 4, pp. 357-363.

Uhrik, C. and Mezgar, I. (1990) Qualitative reasoning as an aid for a learning process that controls a manufacturing line, *Automatic Control in the Service of Mankind, 11th Triennial World Congress, IFAC*, Tallin, USSR, vol. 4, pp. 24-28.

Venkatasubramanian, V. and Chan, K. (1989) A neural network methodology for process fault diagnosis, *AIChE Journal*, 35(12), 1993-2002.

VerDuin, W.H. (1990) Neural networks for diagnosis and control, *J. of Neural Computing*, 1(3), 46-52.

Watanabe, K., Matsuura, I., Abe, M., Kubota, M. and Himmelblau, D.M. (1989) Incipient fault diagnosis of chemical process via artificial neural networks, *AIChE Journal*, 35(11), 1803-1812.

Willis, M.J., Di Massimo, C., Montague, G.A., Tham, M.T. and Morris A.J. (1991) Artificial neural networks in process engineering, *IEE Proc. part D*, 138(3), 256-265.

Winchell, W. and Chugh, N. (1989) An expert system for generic quality improvement, *43rd Annual Quality Congress Transactions*, Toronto, Ontario, 8-10 May 1989, pp. 424-427.

With, I. and Irgens, C. (1990) Knowledge-based support of quality in design, *Computer Integrated Manufacturing*, Floria L. and Puymbroeck V. (eds). Springer Verlag, UK, pp. 245-258.

Wu, H.J., Liou, C.S. and Pi, H.H. (1991) Fault diagnosis of processing damage in injection moulding via neural network approach, *Intelligent Engineering Systems Through Artificial Neural Networks, Proc. of the Artificial Neural Networks in Engineering (ANNIE'91) Conf.*, St Louis, Missouri, 10-13 November 1991, pp. 645-650.

Yih, Y. and Nof, S. (1991) Impact of integrating knowledge-based technologies in manufacturing: An evaluation, *Computer Integrated Manufacturing Systems*, 4(4), 254-263.

Chapter 2   Artificial Intelligence Tools

This chapter reviews expert systems and neural networks, two of the main artificial intelligence tools that have been applied to quality control tasks in manufacturing. The chapter first discusses the basic components of an expert system and briefly surveys expert system development tools including various commercially available shells and environments. The fundamentals of neural networks are then described. Different neural network architectures and learning strategies are detailed. Examples of applications will be presented in subsequent chapters of the book.

2.1 Expert Systems

Expert systems are a product of artificial intelligence (AI), the branch of computer science concerned with developing programs that exhibit "intelligent" behaviour. There is a vast amount of literature describing the principles and features of expert systems. See for example [Pham and Pham, 1988; Martin and Oxman, 1988; Lucas and Van der Gaag, 1991; Hopgood, 1993; Durkin, 1994]. The past decade has witnessed expert systems progress from prototypes to products deployed in industrial settings, performing tasks involving control, debugging, design, diagnosis, interpretation, instruction, monitoring, planning, prediction and repair [Pham, 1988; Pham, 1991; Shamsudin and Dillon, 1991; Tzafestas, 1993; Pham, 1994; Liebowitz, 1996]. A range of development environments or tools and shells for building and maintaining expert systems are commercially available. This greatly facilitates the construction of expert systems and further increases the number and range of practical applications.

2.1.1 Knowledge Representation

The heart of an expert system is the domain knowledge (knowledge about a particular problem or situation). Therefore, expert systems are also referred to as knowledge-based systems. In an expert system, the domain knowledge is usually represented in two forms: it is either at the level of know-how where underlying fundamentals are not detailed (this form of knowledge is known as shallow knowledge), or at a level where its theoretical and scientific fundamentals are deeply expressed (this is called deep knowledge). An analysis of the two forms of knowledge may be found in Price and Lee [1988] and Lavangnananda [1995]. There are several ways of representing either type of knowledge in an expert system. The three most popular methods, rules, frames and semantic networks, are reviewed here.

Rule-based knowledge comprises the essential facts pertinent to a problem and a set of rules for manipulating the facts. Facts are usually asserted in the form of statements which explicitly classify objects or show their relationships. For example, facts may be simple statements like: 'A Mercedes is a Car', 'A Car requires an Engine' or 'An Engine generates Mechanical Power' whereas rules are modular pieces of knowledge of the form:-

IF antecedent THEN consequent

or

IF condition THEN action

That is, if the stated condition is satisfied then the specified action is performed. These rules are called IF/THEN rules, situation-action rules or production rules.

In contrast to rules which express shallow knowledge, representation schemes using frames or semantic nets allow a deeper insight into the underlying concepts and causal relationships, facilitating the implementation of deeper level reasoning. A frame is a record-like data structure, a form for encoding information on a stereotyped situation, a class of objects, a general concept or a specific instance of any of these. Associated with each frame is a set of attributes, the descriptions or values of which are contained in the slots. For example, a frame for the 'Mercedes' instance of the class of 'Car' objects could have slots entitled 'colour', 'owner', 'price' and 'year of registration'.

Knowledge representation schemes using semantic networks are similar to those based on frames. A semantic net is a network of nodes linked together by arcs. Nodes stand for general concepts (types), specific objects (tokens), general events (prototyped events) or specific events. Arcs describe relations (such as 'is-a' or 'has-part' relations) between nodes.

2.1.2 Elements of an Expert System

The main components of an expert system, as illustrated in Figure 2.1, are briefly explained in this section.

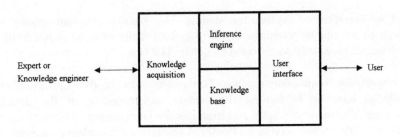

Figure 2.1 Main components of an expert system

(i) **Knowledge base.** This contains knowledge about the problem domain. It can comprise rules, rulesets, frames, classes and procedures. Rules, frames and classes have already been explained. Rulesets are groups of related rules designed to facilitate knowledge management. Procedures are sets of instructions for performing routine algorithmic tasks. Examples of rules, rulesets, frames, classes and procedures for a particular problem are given in Chapter 3. An expert system, where the knowledge base consists of rules and rulesets, is known as a rule-based expert system.

(ii) **Inference engine.** This manipulates the stored knowledge to produce solutions to problems. The inference engine in a rule-based expert system scans the knowledge base, selecting and applying appropriate rules. Inferencing can proceed in different ways according to different control procedures. One strategy is to start with a set of facts or given data and look for those rules in the knowledge base whose **IF** part matches the given fact/data. When such rules are found, one of them is selected based upon an appropriate *conflict resolution* criterion and executed or *fired*. This generates new facts and data in the knowledge base which in turn causes other rules to be executed. This kind of reasoning is known as *forward chaining* or *data-driven inferencing*. An alternative approach is to begin with the goal to be proved and try to establish the facts needed to prove it by examining the **THEN** part of rules with the desired goal. If such facts are not available in the knowledge base, they are set up as sub-goals which are in turn to be proved. This method of reasoning is called *backward chaining* or *goal-directed inferencing*. In practice, these two methods of reasoning are sometimes combined to give a more efficient problem solving method.

(iii) **User interface** or **explanation module.** This handles communication with the user in a "natural" language. A set of general facilities to be provided by a user interface module is documented by Zahedi [1990].

(iv) **Knowledge acquisition module.** This assists with the development of the knowledge base by facilitating the capture and encoding of the domain knowledge. The main principles and strategies for knowledge acquisition may be found in Cullen and Bryman [1988]. A promising knowledge acquisition technique is machine learning by induction from examples. Simple and efficient induction algorithms have been developed for this purpose. For instance, see Pham and Aksoy [1995].

2.1.3 Expert System Shells and Tools

A number of languages and development tools and shells are available to build knowledge-based or expert systems. Tools and shells are not as flexible as high-level languages but are easier to use for expert system development. Since the number of tools is large, choosing the best for a particular problem is a difficult task. Guidelines on how to select an appropriate tool and a cost-benefit analysis of commercially available tools are given by Mackerle [1989], Price [1990] and Mettrey [1991].

Expert system building tools can be divided according to their size (capabilities) into development environments tools and shells (see Figure 2.2). Development environments can be further categorised based on the range of knowledge representation tools which they provide. Shells, on the other hand, offer a relatively restricted set of facilities and are usually less expensive.

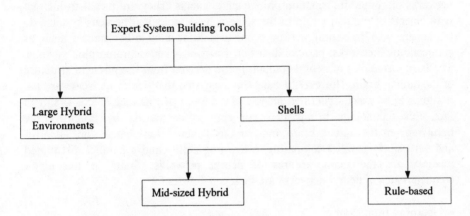

Figure 2.2 Classification of expert system tools

Development environments are a collection of various tools and represent the state-of-the-art in expert system technology. Such tools are used by experienced knowledge engineers who act as the interpreter between the domain expert and the hardware/software environment. These tools can be implemented on mainframes, workstations and advanced personal computers. Three well known development environments are ART, KEE and KAPPA.

Shells are implemented mainly on microcomputers. Some shells may be considered mid-sized hybrid tools having more than one knowledge representation paradigm. A number of commercial products are available in this category, for example, KAPPA-PC, NEXPERT Object and Gold Works. However, there are shells which provide only one knowledge representation paradigm, such as Crystal, Xi Plus and Super Expert. The paradigm usually adopted in these shells is rule-based representation.

2.2 Neural Networks

Research in the last decade has shown that neural networks are useful tools for solving many practical engineering problems [Roth, 1988; Stevenson, 1991; Pham and Liu, 1995]. Neural networks are highly parallel interconnected networks of simple information processing elements. They are meant to interact with objects of the real world in the same way as biological nervous systems do [Kohonen, 1987]. Neural networks are also known by other names such as connectionist networks, parallel distributed networks and neuromorphic systems. The basic principles of neural computing are derived from the physical structure of the human brain. However, it has been suggested that neural networks are not designed to be exact **structural** models of the brain [Karna and Breen, 1989]. In fact, this technology brings new concepts to computing by imitating the **behaviour** of the human brain and nervous system. Non-algorithmic, adaptive and extensively parallel computing, a learning ability and a parallel distributed memory are the main features of neural networks. Some of the major characteristics of neural networks are their capabilities for:-

- learning from examples;
- pattern association and classification;
- pattern auto-association (reconstruction of an incomplete pattern);
- self-organisation and learning (adaptability);
- fault tolerance (a neural network does not completely fail in the case of destruction or removal of some neurons);
- immunity to fuzzy or noisy inputs.

2.2.1 Fundamentals of Neural Networks

A general neural network is composed of a number of neurons (also called processing elements). Each neuron consists of five components: inputs, weights, a combination function, an activation function and an output (see Figure 2.3). The information to each neuron derives either from an external source or from other neurons. In some cases, the neuron can also feed information back to itself. Each weight determines the influence of an input on the neuron and can be variable or fixed. The combination function defines the net input to the neuron which is usually the weighted *sum* of all inputs. Other combination functions include *maximum, minimum, majority* and *product* (see Figure 2.4). The activation function is applied to the result of the combination function to yield the neuron output. Some example activation functions are given in Figure 2.5. The output is sent to other neurons or to an external system. For each neuron there may be multiple inputs but only one output. A neuron may function independently or in parallel with other neurons in the network.

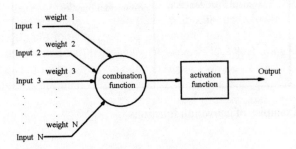

Figure 2.3 Structure of a neuron (processing element)

Sum : $\text{Net}_i = \sum_j W_{ij} I_j$	Product : $\text{Net}_i = \prod_j W_{ij} I_j$
Maximum : $\text{Net}_i = \text{Max}(W_{ij} I_j)$	Minimum : $\text{Net}_i = \text{Min}(W_{ij} I_j)$
Majority : $\text{Net}_i = \sum_j \text{sgn}(W_{ij} I_j)$	Cumulative Sum : $\text{Net}_{new} = \text{Net}_{old} + \sum_j W_{ij} I_j$

i, j : neuron
Net_i : net input to neuron i
I_j : output of neuron j
W_{ij} : weight of connection between neurons j and i
sgn : signum function

Figure 2.4 Examples of combination functions

Linear function : f(x) = x	Step function : $f(x) = \begin{cases} 1 & \text{if } x > \text{threshold} \\ 0 & \text{if } x <= \text{threshold} \end{cases}$
Sigmoid function : $f(x) = \dfrac{1}{1 + e^{-x}}$	Hyperbolic tangent function : $f(x) = \dfrac{e^x - e^{-x}}{e^x + e^{-x}}$
Threshold logic function : $f(x) = \begin{cases} 0 & \text{if } x <= 0 \\ x & \text{if } 0 < x < 1 \\ 1 & \text{if } x >= 1 \end{cases}$	Sinusoidal function : $f(x) = \sin(x)$

Figure 2.5 Examples of activation functions

The neurons may be *locally* connected to their neighbours, *fully* connected to all other neurons or *sparsely* connected to a few distant neurons [DARPA, 1988]. Connections can be unidirectional (feedforward) or bi-directional (recursive). A group of neurons can form a structure called a *layer*. A typical network consists of an *input layer* which receives inputs from the external world, one or more *hidden (intermediate) layers* which are responsible for the main information processing and an *output layer* which delivers the response of the network to the external world. A neural network usually functions by accepting a pattern in its input layer, propagating that through its hidden layers and producing an output pattern at the output layer.

Information in a neural network is represented by the weights of the connections between neurons and is distributed throughout the network. It is retrieved through association. That is, when a pattern is presented to the network, an associated output is produced. This output pattern represents the actual information held by the network about the input pattern.

Although most neural network models are based on the above description, many different topologies have been proposed. These are usually characterised by the combination function, activation function, network topology and learning rule that they employ [Masson and Wang, 1990]. Details of the different models can be found in the literature, for example [Lippmann, 1987; Humpert, 1990; Dayhoff, 1990; Simpson, 1990, Pham and Liu, 1995]. Three of the most popular among the available neural network models, the Multi-Layer Perceptron (MLP), Learning Vector Quantisation (LVQ) and Adaptive Resonance Theory networks (ART), are described in this chapter. Examples of the use of these models for quality control will be given in later chapters.

2.2.2 Learning

The "intelligence" of a neural network is contained in the strengths of the connections (weights) between its neurons. Since knowledge is distributed throughout the network, a single connection weight does not by itself correspond to any meaningful information. Rather, all the connections together constitute meaningful knowledge. It should be noted that a network must have suitable weights for all connections in order to display intelligent behaviour. These weights are assigned by applying a learning rule. The latter is an algorithm which determines how a neuron will change its connection weights through experience. Several learning rules, using different learning strategies, have been developed [Simpson, 1990]. These include the generalised delta rule and the Kohonen learning rule which are described in the following sections. Learning rules are grouped into learning strategies. There are three learning strategies: supervised learning, reinforcement learning and unsupervised learning.

The **supervised learning** strategy [Rumelhart *et al.*, 1986a] requires an external teacher who knows the desired output corresponding to a particular input pattern. The network is given an input-output pair which is an example of the mapping it is supposed to learn. The learning process is viewed as a means to help the network filter the input pattern, produce an output pattern and compare that with the desired output pattern. Weights are then modified by a learning rule to reduce the error between the actual and desired outputs.

The **reinforcement learning** strategy [Carling, 1992] also requires a teacher. However, in this case, the network is just given a reinforcement signal telling it whether the output produced is good or bad without giving any details of the output pattern itself. This strategy is useful when it is difficult to form the desired outputs corresponding to the input patterns.

The **unsupervised learning** strategy [Carpenter and Grossberg, 1987; Kohonen, 1987] does not require a teacher and is sometimes called self-organising learning. The network categorises the input patterns into subsets according to predetermined similarity criteria. The only information available to the network is the set of input patterns presented to it. The network must extract knowledge from this information and develop its own classification rules.

2.2.3 Architecture of a Multi-Layer Perceptron (MLP)

The popularisation of the backpropagation learning scheme for the Multi-Layer Perceptron by Rumelhart *et al*. [1986b] has played a large part in the resurgence of interest in neural networks. The scheme expanded the range of problems to which neural networks can be applied, which now include various input-output mapping tasks for recognition, classification and generalisation. Backpropagation learning in a MLP model involves using an iterative gradient descent algorithm to minimise the mean square error between the actual outputs of the network and the desired outputs in response to given inputs.

Network Topology. Figure 2.6 shows the topology of a Multi-Layer Perceptron. It is a feedforward network consisting of neurons in an input layer, one or several hidden (intermediate) layers and an output layer. Each neuron in one layer is fully connected to all neurons in the next layer. There are no connections between neurons in the same layer or back to neurons in previous layers. The input layer, which is also called the "buffer" layer, performs no information processing. Each of its neurons has only one input and simply transmits the value at its input to its output. Actual information processing is performed by the neurons in the hidden and output layers. Information flows from the input layer to the output layer through the hidden layers and errors between the actual and desired outputs are propagated in the reverse direction.

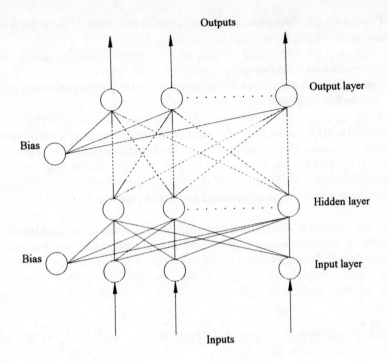

Figure 2.6 Topology of a MLP

MLP Procedure. The following steps are involved in constructing and teaching a MLP network:-

(i) Defining the structure of the network (the number of layers and the number of neurons in each layer);
(ii) Selecting learning parameters (the learning rate and momentum coefficient which are defined in the next section);
(iii) Initialising the connection weights;
(iv) Selecting an input-output pair from the training example set and presenting it to the network;
(v) Calculating the values of the outputs of the neurons in the hidden and output layers;

(vi) Comparing the calculated output values with the desired output values and computing the output errors;

(vii) Adjusting the connection weights of the network according to the output errors in order to decrease them;

(viii) Repeating (iv) to (vii) until the error is acceptable or a predefined number of iterations are completed.

Learning in MLP networks. As mentioned above, training a MLP network involves forward and backward operations. In the forward operation, the network produces its outputs for a given input pattern using the current connection weights. Subsequently, the backward operation is carried out to alter the weights to decrease the error between the actual and desired outputs.

Forward Processing: This operation starts with the input layer. A pattern from the training set is presented to the network and conveyed to the hidden layer without change. That is:-

$$OUT_k^i = I_k \tag{2.1}$$

where I_k is the value of attribute k in input pattern i and OUT_k^i is the output value of neuron k in the input layer.

Each neuron j in hidden layer h accepts the output values of the input layer neurons and computes its net input NET_j^h as:-

$$NET_j^h = \sum_k W_{kj}^{i,h} OUT_k^i \tag{2.2}$$

where $W_{kj}^{i,h}$ is the weight of the connection between neurons k and j in the input and hidden layers respectively.

The output value of the jth neuron OUT_j^h in the hidden layer is then calculated by passing its net input through an activation function, usually a sigmoidal function. That is:-

$$OUT_j^h = \frac{1}{1 + e^{-(NET_j^h + \beta_j)}} \tag{2.3}$$

where β_j is the bias of the jth neuron. The bias is derived from a special neuron with a fixed activation value, set equal to 1. The output of the bias neuron is connected through links with variable weights to all neurons of the network apart from those in the input layer. The bias offsets the origin of the activation function to improve learning.

The net input and output values for neurons in subsequent layers are calculated in the same way. Once the output values for the output layer neurons are calculated, forward processing is completed.

Backward Processing: The actual output values for the network are compared with the desired output values corresponding to the given input pattern. The difference E_m between these output values for neuron m is:-

$$E_m = (D_m - OUT_m^o) \tag{2.4}$$

where D_m and OUT_m^o are respectively the desired and actual output values for neuron m in output layer o.

Given the current set of weights, the aim is to determine how to modify them to decrease the global sum squared error G_e defined as:-

$$G_e = \frac{1}{2} \sum_m E_m^2 \tag{2.5}$$

According to the backpropagation algorithm [Rumelhart *et al.*, 1986b], the amount of change in the connection weights in iteration $(n+1)$ is given by:-

$$\Delta W_{jm}^{h,o}(n+1) = \lambda \delta_m^o OUT_j^h + \alpha \Delta W_{jm}^{h,o}(n) \tag{2.6}$$

for the weights of the connections between the hidden and output layers, and

$$\Delta W_{kj}^{h-1,h}(n+1) = \lambda \delta_j^h OUT_k^{h-1} + \alpha \Delta W_{kj}^{h-1,h}(n) \tag{2.7}$$

for the weights of the connections between the input and hidden layers or between two hidden layers, where:-

$$\delta_m^o = OUT_m^o (1 - OUT_m^o) E_m \tag{2.8}$$

$$\delta_j^h = OUT_j^h (1 - OUT_j^h) (\sum_m \delta_m^{h+1} W_{jm}^{h,h+1}) \tag{2.9}$$

λ is the learning rate;
α is the momentum coefficient;
n is the iteration number;
h-1 is the previous layer index;
h+1 is the next layer index.

The learning rate defines the range of the changes in the connection weights, the larger the learning rate the larger the changes in the connection weights. The momentum coefficient effectively adds a term to the weight adjustment that is proportional to the previous weight change. This helps to dampen oscillations and accelerate learning [Lippmann, 1987].

The new connection weights are then given by:-

$$W(n+1) = W(n) + \Delta W(n+1) \tag{2.10}$$

2.2.4 Architecture of a Learning Vector Quantisation (LVQ) Network

The LVQ model has been developed by Kohonen [1984]. As its name indicates, it is based on vector quantisation which is the mapping of an n-dimensional vector into one belonging to a finite set of representative vectors [Gersho, 1982]. That is, vector quantisation involves clustering input samples around a predetermined number of reference vectors. Learning in a LVQ network essentially consists of finding those reference vectors.

The classification of input vectors into clusters is conducted on the basis of the nearest neighbour rule. The smallest distance between the input vector and reference vectors is sought. Reinforcement learning, with the winner-takes-all learning strategy, is adopted. This means that, at each learning iteration, the network is only told whether its output is correct or incorrect and only the reference vector of that neuron which wins the competition, the one closest to the input vector, is activated and has its connection weights modified. The general features of LVQ are shown in Figure 2.7.

Learning Vector Quantisation (LVQ)

- Vector quantisation
- Nearest-neighbour classifier
- Reinforcement learning
- Winner-takes-all learning strategy
- Monotonically decreasing learning rate

Input layer	Hidden layer	Output layer
Fully connected	Partially connected	

Figure 2.7 Features of a LVQ network

Network Topology. A LVQ network is composed of three layers: an input layer which does no information processing and only conveys the input patterns to the network as in the MLP case, a hidden layer (also known as the Kohonen layer) which performs information processing and an output layer which yields the category of the input pattern. The network is fully connected between the input and hidden layers and partially connected between the hidden and output layers (see Figure 2.8). The hidden layer to output layer connections have their values fixed at 1. The weights of the connections between the input and hidden layers constitute the components of the reference vectors. Their values are modified during learning. The neurons in the hidden and output layers (Kohonen and output neurons) produce binary output values.

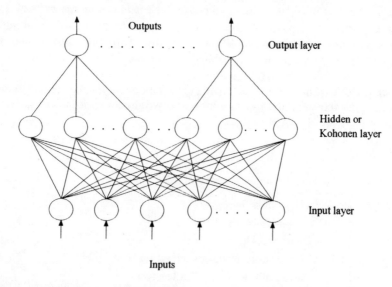

Figure 2.8 Topology of a LVQ network

When a Kohonen neuron wins the competition, it is turned "on" (its activation value is made equal to 1) while others are automatically switched "off" (their activation values are set to 0). This, in turn, makes the output neuron connected to the activated Kohonen neuron switch "on" and the rest switch "off".

LVQ Procedure. The following steps are involved in constructing and teaching a LVQ network:-

(i) Defining the network structure (deciding on the number of inputs and outputs, the number of reference vectors for each output category and the learning rate);
(ii) Initialising the weights of the reference vectors;
(iii) Presenting an input vector from the training set;
(iv) Calculating the distances between the input vector and the reference vectors;

(v) Updating the nearest reference vector according to some learning rule;
(vi) Repeating (iii) to (v) until all patterns are correctly classified or the required number of iterations has been completed (stopping criterion).

Good performance in a LVQ network depends on the correct number of reference vectors being assigned to each category, their initial values and the choice of a proper learning rate and stopping criterion.

Standard LVQ Rule. A LVQ network is trained by enabling competition to take place between the Kohonen neurons. The competition is based on the Euclidean distances between the weight vectors of these neurons (the reference vectors) and the input vector. The distance d_i between the weight vector W_i of neuron i and the input vector X is given by:-

$$d_i = \left\| W_i - X \right\| = \sqrt{\sum_j (W_{ij} - X_j)^2}$$

(2.11)

where W_{ij} and X_j are the jth components of W_i and X respectively.

As already mentioned, the neuron that has the minimum distance wins the competition and is allowed to change its connection weights. The weights of the other neurons remain unchanged. The new weights are given by:-

$$W_{new} = W_{old} + \lambda (X - W_{old})$$

(2.12)

if the winning neuron is in the correct output category (that is the category which the input vector is known to belong to), or

$$W_{new} = W_{old} - \lambda (X - W_{old})$$

(2.13)

if the winning neuron is in the wrong category.

In the above equations, λ is the learning rate which decreases monotonically with the number of iterations. The implication of the learning rule expressed in these equations is that the reference vector should be pulled closer to the input vector if it represents the input pattern and pushed away if it does not.

Variants of LVQ Learning Rule. There are a number of LVQ rules which have been developed to improve the efficiency of the standard model explained above. LVQ2 and LVQ with a *conscience mechanism* are two of the variants of the original LVQ model.

LVQ2: LVQ2 has also been developed by Kohonen [Kohonen *et al.*, 1988]. It is usually employed after acceptable results have been obtained by applying the standard procedure. LVQ2 refines the solution boundary between regions where misclassifications have occurred. The learning algorithm simultaneously modifies two reference vectors W_1 and W_2 in each learning iteration if:-

(i) W_1 and W_2 are the closest and next closest neighbours of the input vector, W_1 is in the incorrect category and W_2 is in the correct category, and
(ii) the input vector X falls into a window centrally located between W_1 and W_2. The input vector is said to be inside such a window when [Kohonen, 1990]:-

$$\min (d_1/d_2, d_2/d_1) > s \qquad (2.14)$$

where:-

$s = (1-w)/(1+w)$
d_1 is the distance between the input vector and W_1;
d_2 is the distance between the input vector and W_2;
w is the window width.

The new reference vectors $W_{1\ new}$ and $W_{2\ new}$ are given by:-

$$\mathbf{W}_{1\ new} = \mathbf{W}_{1\ old} - \lambda(\mathbf{X} - \mathbf{W}_{1\ old}) \qquad (2.15)$$

$$\mathbf{W}_{2\ new} = \mathbf{W}_{2\ old} + \lambda(\mathbf{X} - \mathbf{W}_{2\ old}) \qquad (2.16)$$

Other connection weights of the network remain untouched. This weight modification process is represented geometrically in Figure 2.9.

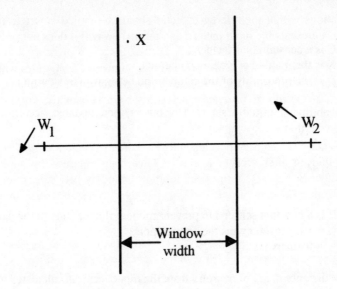

Figure 2.9 Geometric representation of the LVQ2 procedure

LVQ with a Conscience Mechanism: The standard Learning Vector Quantisation algorithm can suffer from the problem that some neurons tend to win too often while others are always inactive. This particularly happens when the neurons are initialised far from the training vectors. In this case, some neurons would quickly move closer to the training vectors and the others remain permanently far away. To help prevent this, a "conscience" mechanism has been suggested by DeSieno [1988]. This mechanism gives the neuron which wins too often a "guilty conscience", penalising it by adding a distance bias to the true distance between that neuron and the input vector. The distance bias is based on the number of times the neuron has won the competition. It is calculated as:-

$$b_i = C\left(p_i - \frac{1}{N}\right) \qquad (2.17)$$

where:-

C is a constant bias factor;

N is the number of Kohonen neurons;

p_i is the probability of the competition being won by neuron i.

This probability is initially set at 1/N but is then updated according to the following equation:-

$$p_{i\,new} = p_{i\,old} + B(y_i - p_{i\,old})$$ (2.18)

where:-

B is a constant selected to prevent random fluctuations in the data;

y_i = 1 if neuron i wins the competition;

= 0 otherwise.

The new distance $d_{i\,new}$ of neuron i from the input vector is calculated as:-

$$d_{i\,new} = d_{i\,old} + b_i$$ (2.19)

The competition is carried out with the new distances and the weight updating rule for the standard LVQ is then applied.

An extended version of the LVQ learning rule, called LVQ-X, developed to recognise control chart patterns will be explained in Chapter 4.

2.2.5 Architecture of an ART2 Network

The ART2 network has been developed by Carpenter and Grossberg [1987]. It is an improvement over the ART1 network which is only able to cope with binary inputs [Carpenter and Grossberg, 1985]. The ART2 paradigm is able to handle both binary and analogue input patterns. ART2 networks are self-organising networks which match and learn sequences of input patterns in a stable fashion without requiring supervision.

Network Topology. The general architecture of an ART2 network is given in Figure 2.10. This structure basically contains an attentional subsystem comprising two layers (F1 and F2) and an orienting subsystem. F1 is usually referred to as the feature representation field and F2 as the category representation field. In the F1 layer, there are two types of neuron populations: the linear neurons such as p_i, v_i, and w_i, and the shunting neurons such as q_i, x_i, and u_i. Linear neurons simply add up their inputs. Shunting neurons normalise their inputs. The connections between the two layers F1 and F2 are named long term memory (LTM) traces and constitute the bottom-up and the top-down filter. The outputs of the neurons are called the short term memory (STM) traces. The network has short term memory traces in both layers F1 and F2. It is important to note that the orienting subsystem is only activated when a bottom-up input to F1 fails to match with the established top-down expectation read-out by the active category in layer F2.

ART2 Procedure. The following steps are involved in training an ART2 network:-

(i) Initialising the top-down and bottom-up long term memory traces;

(ii) Presenting an input pattern from the training data set to the network;

(iii) Triggering the neuron with the highest total input in the category representation field according to the mechanism of lateral inhibition (competitive learning);

(iv) Checking the match between the input pattern and the exemplar in the top-down filter (long term memory) using a vigilance parameter;

(v) Starting the learning process if the mismatch is within the tolerance level defined by the vigilance parameter and then going to step (viii), otherwise activating the orienting subsystem and moving to the next step;

(vi) Disabling the current active neuron in the category representation field and returning to step (iii). Going to step (vii) if all the established classes have been tried;

(vii) Establishing a new class for the given input pattern and restarting the learning process;

(viii) Repeating (ii) to (vii) until the network stabilises or a specified number of iterations are completed.

Note that during the recall mode, only steps (ii),(iii), (iv) and (viii) will be utilised.

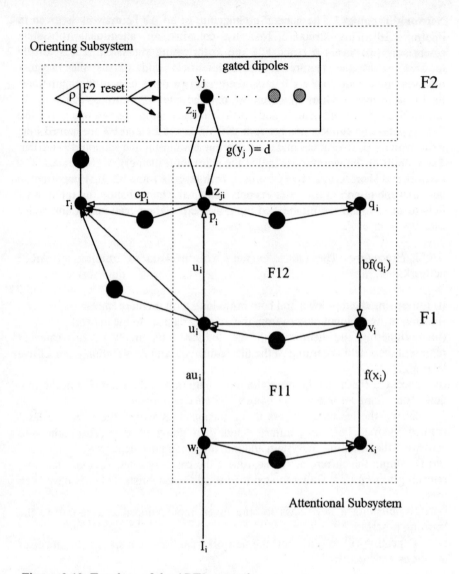

Figure 2.10 Topology of the ART2 network

Learning in an ART2 Network. Learning in an ART2 network depends on a procedure called hypothesis testing. When the network is provided with a new input pattern, the network checks if it is possible to classify the input pattern into one of the established classes. If it is possible, then learning starts (the attentional subsystem is activated). If not, the network creates a new class for that particular input pattern (the orienting subsystem is activated) and then learning starts.

The network encodes new input patterns, in part, by changing the weights or long term memory (LTM) traces of the top-down adaptive filter between the feature and the category representation fields. The second filter, which is also important in learning, is the bottom-up filter that gives the crucial property of code self-stabilisation and speedy classification. This filter enables the network to carry out attentional priming, pattern matching and self-adjusting parallel search.

Short Term Memory Module in the F1 Layer: The aim of this module is to try to recirculate, temporarily, a pattern from the input layer and an expected top-down pattern stored in a higher layer. This implements a short term memory where the energy of neural activity represents the correlation between the patterns. This means that in this layer, the contrast between the filtered input pattern and the expected top-down pattern is enhanced.

The potential or STM activity of neuron V_i at any one of the processing stages obeys the following membrane equation:-

$$ e \frac{d}{dt} V_i = -AV_i + (1-BV_i)J_i^+ - (C+DV_i)J_i^- \tag{2.20} $$

where J_i^+ is the total excitatory input to the ith neuron, J_i^- is the total inhibitory input and e is the ratio between the STM relaxation time and the LTM relaxation time. In the architecture given in Figure 2.10, the neurons obey a simplified equation by considering that B=0 and C=0. If the processing in the STM neurons takes place very quickly, these neurons can be regarded as being in a steady state. This means that the equation becomes:-

$$V_i = \frac{J_i^+}{A + DJ_i^-} \tag{2.21}$$

Considering DJ_i^- as the norm of pre-synaptic neurons in the circuit of layer F1, inhibitory inputs to a shunting neuron lead to normalisation. The dimensionless parameters p_i, q_i, u_i, v_i, x_i, which characterise the STM activities in layer F1, are given by:-

$$p_i = u_i + \sum_j g(y_j) z_{ij} \tag{2.22}$$

$$q_i = \frac{p_i}{e + \|\mathbf{p}\|} \tag{2.23}$$

$$u_i = \frac{v_i}{e + \|\mathbf{v}\|} \tag{2.24}$$

$$v_i = f(x_i) + bf(q_i) \tag{2.25}$$

$$w_i = I_i + au_i \tag{2.26}$$

$$x_i = \frac{w_i}{e + \|\mathbf{w}\|} \tag{2.27}$$

where:-

$\|\ \|$ denotes the L_2-norm of a vector,

I is the input pattern,

y_j is the STM activity of the jth neuron in F2,

f is a non-linear function which is usually defined for continuously differentiable cases as

$$f(x) = \begin{cases} \dfrac{2\pi x^2}{x^2 + \tau^2} & \text{if } 0 \le x \le \tau \\[4mm] x & \text{if } x \ge \tau \end{cases} \qquad (2.28)$$

and for piecewise linear cases as

$$f(x) = \begin{cases} 0 & \text{if } 0 \le x < \tau \\ x & \text{if } x \ge \tau \end{cases} \qquad (2.29)$$

$g(y)$ is a winner-takes-all function which permits only the most active F2 layer neuron to fire,
a and b are gain parameters which are set by the user.

Short Term Memory Module in the F2 Layer: The key functions of the F2 layer are contrast enhancing filtered input patterns received from the F1 layer and resetting the active F2 nodes whenever a mismatch is large enough to activate the orienting subsystem. Contrast enhancement is carried out by competition within the F2 layer. The neuron receiving the largest total input is selected (winner-takes-all strategy). In other words, let T_j be an output of the bottom-up filter and the input to the jth neuron in layer F2.

$$T_j = \mathbf{p} \cdot \mathbf{z_j} = \sum_i p_i z_{ij} \qquad \text{for } j = M + 1, \ldots\ldots, N \qquad (2.30)$$

Here, M is the dimension of the input vector (the number of neurons in each population in the F1 layer) and N-M is the number of available neurons in the F2 layer. For the winning neuron, neuron J, T_J is given as :-

$$T_J = \text{Max}[T_j : j = M + 1, \ldots\ldots, N] \qquad (2.31)$$

The F2 internal reset mechanism can be easily implemented using a gated dipole field network. When a non-specific arousal input arrives, the nodes in the F2 gated dipole field are inhibited or reset proportionally to their previous STM activity levels. This inhibition lasts only until the bottom-up input to F1 shuts off. Such a non-specific arousal wave reaches F2 via the orienting subsystem when a sufficiently large mismatch occurs in F1 during the selection of the Jth neuron in F2. The gated dipole field dynamics can be characterised as:-

$$g(y_j) = \begin{cases} d & \text{for the selected neuron} \quad (0 < d < 1) \\ 0 & \text{otherwise} \end{cases} \tag{2.32}$$

Long Term Memory Module in the F2 Layer: The heart of the ART2 architecture is two sets of adaptive filters or long term memory traces. Each category is coded with a bottom-up and top-down filter. During training, the filter coefficients are modified to such an extent that the filter vector is nearly aligned with the input pattern vector. The top-down LTM trace equation is:-

$$F2 \rightarrow F1 \quad \frac{d}{dt} z_{ji} = g(y_i)[p_i - z_{ji}] \tag{2.33}$$

The bottom-up equation is also given in the same way as:-

$$F1 \rightarrow F2 \quad \frac{d}{dt} z_{ij} = g(y_i)[p_i - z_{ij}] \tag{2.34}$$

If the F2 layer selects node J as the winner and it is still active then the above equation becomes:-

$$\frac{d}{dt} z_{Ji} = d(p_i - z_{Ji}) = d(1-d)\left[\frac{u_i}{1-d} - z_{Ji}\right] \tag{2.35}$$

$$\frac{d}{dt} z_{iJ} = d(p_i - z_{iJ}) = d(1-d)\left[\frac{u_i}{1-d} - z_{iJ}\right] \tag{2.36}$$

with $0<d<1$. For all other neurons, which lose the competition, the above equations are set to 0.

Orienting Subsystem: The match between a STM pattern in F1 and an active LTM pattern is numerically expressed by a vector denoted as **r**. The components of this vector are calculated as:-

$$r_i = \frac{u_i + cp_i}{e + \|\mathbf{u}\|\|\mathbf{p}\|} \tag{2.37}$$

where c is a gain parameter set by the user.

The orienting module is assumed to reset F2 whenever an input pattern is active and

$$\frac{\rho}{e + \|\mathbf{r}\|} > 1 \tag{2.38}$$

where ρ is a vigilance parameter set by the user between 0 and 1.

In computer simulations, for simplicity, e can be set to 0. Since x_i, u_i and q_i are shunting neurons responsible for normalising their inputs, $\|\mathbf{x}\|$, $\|\mathbf{u}\|$ and $\|\mathbf{q}\|$ are all equal to 1.

2.3 Summary

This chapter has described two of the artificial intelligence techniques which have been applied to several quality control tasks in manufacturing. The chapter has first provided an overview of expert system technologies and then discussed the basic components of an expert system, giving a synopsis of expert system development tools. The chapter has also examined neural networks, detailing three different neural network models representing the three main learning strategies, the MLP model (supervised learning), the LVQ model (reinforcement learning) and the ART2 model (unsupervised learning).

References

Carling, A. (1992) *Introducing Neural Networks*, Sigma Press, UK.

Carpenter, G.A. and Grossberg, S. (1985) Category learning and adaptive pattern recognition: A neural network model, *Proc. of the 3rd Army Conference on Applied Mathematics and Computing*, ARO Report 86-1, pp. 37-56.

Carpenter, G.A. and Grossberg, S. (1987) ART2: Self-organization of stable category recognition codes for analog input patterns, *Applied Optics*, 26(23), 4919-4930.

Cullen, J. and Bryman, A. (1988) The knowledge acquisition bottleneck: Time for reassessment, *Expert Systems*, 5(3), 216-225.

DARPA (1988) *Neural Network Study*, Fairfax, VA: AFCEA International, pp. 411-415.

Dayhoff, J.E. (1990) *Neural Network Architectures*, Van Nostrand Reinhold, New York.

DeSieno, D. (1988) Adding a conscience to competitive learning, *Proc. of the Int. Joint Conf. on Neural Networks*, San Diego, California, vol. 1, pp. I-117 - I-124.

Durkin, J. (1994) *Expert Systems Design and Development*, Macmillan, New York, NY.

Gersho, A. (1982) On the structure of vector quantizers, *IEEE Transactions on Information Theory*, IT-28(2), 157-166.

Hopgood, A.A. (1993) *Knowledge-Based Systems for Engineers and Scientists*, CRC Press, Boca Raton, FL.

Humpert, B. (1990) A comparative study of neural network architectures, *Computer Physics Communications*, vol. 58, 223-256.

Karna, K.N. and Breen, D.M. (1989) An artificial neural networks tutorial: Part 1 - Basics, *Neural Networks*, 1(1), 4-23.

Kohonen, T. (1984) *Self-Organization and Associative Memory*, Springer Verlag, New York.

Kohonen, T. (1987) State of the art in neural computing, *Proc. of the IEEE First Int. Conf. on Neural Networks*, San Diego, 21-24 June 1987, California, vol. 1, pp.77-91.

Kohonen, T., Barna, G. and Chrisley, R. (1988) Statistical pattern recognition with neural networks: Benchmarking studies, *Proc. of the Int. Joint Conf. on Neural Networks*, San Diego, California, vol. 1, pp. I-61 - I-68.

Kohonen, T. (1990) Self-organizing feature map, *Proc. of the IEEE*, 78(9), 1464-1480.

Lavangnananda, K. (1995) *A Framework for Qualitative Model-Based Reasoning about Mechanisms*, University of Wales, College of Cardiff, UK.

Liebowitz, J. (ed.) (1996) *Proc. of the 3rd World Congress on Expert Systems*, Seoul, Korea, February 1996.

Lippmann, R.P. (1987) An introduction to computing with neural nets, *IEEE Acoustics, Speech and Signal Processing Magazine*, 4(2), 4-22.

Lucas, P. and Van der Gaag, L. (1991) *Principles of Expert Systems*, Addison Wesley, Reading, MA.

Mackerle, J. (1989) A review of expert system development tools, *Engineering Computing*, 6(3), 2-17.

Martin, J. and Oxman, S. (1988) *Building Expert Systems: A Tutorial*, Prentice Hall, Englewood Cliffs, N.J.

Masson, E. and Wang. Y.J. (1990) Introduction to computation and learning in artificial neural networks, *European J. of Operational Research*, vol. 47, pp. 1-28.

Mettrey, W. (1991) A cooperative evaluation of expert system tools, *Computer*, 24(2), 19-31.

Pham, D.T. (1988) *Expert Systems in Engineering*, IFS and Springer Verlag, Berlin.

Pham, D.T. (1991) *Artificial Intelligence in Design*, Springer Verlag, Berlin.

Pham, D.T. (1994) Artificial intelligence in manufacturing, *Proc. Conf on Applications of Artificial Intelligence Techniques in Engineering*, invited lecture, Naples, Italy, October 1994, Chianese, A., Sansone, L., Teti, R. and Zollo, G. (eds.), Liguori Editore, Naples, 5-37.

Pham, D.T. and Liu, X. (1995) *Neural Networks for Identification, Prediction and Control*, Springer Verlag, Berlin and London.

Pham, D. T. and Aksoy, M.S. (1995) A new algorithm for inductive learning, *J. of Systems Engineering*, 5(2), 115-122.

Pham, D.T. and Pham, P.T.N. (1988) Expert systems in mechanical and manufacturing engineering, *Int. J. of Advanced Manufacturing Technology*, 3(3), 3-21.

Price, C.J. (1990) *Knowledge Engineering Toolkits*, Ellis Horwood, Chichester, England.

Price, C. J. and Lee, M. (1988) *Deep Knowledge Tutorial and Bibliography*, Alvey Report IKBS3/26/048, Artificial Intelligence and Robotics Group, University of Wales, Aberystwyth, UK.

Roth, M.W. (1988) Neural network technology and its applications, *John Hopkins Appl. Tech. Dig.*, 9(3), 242-253.

Rumelhart, D.E., Hinton, G.E. and Williams, R.J. (1986a) Learning representations by backpropagating errors, *Nature*, vol. 323, pp. 533-536.

Rumelhart, D.E., Hinton, G.E. and Williams, R.J. (1986b) Learning internal representations by error propagation, *Parallel Distributed Processing*, Rumelhart, D. E. and McClelland, J. L. (eds.), Cambridge MA: MIT press, vol. 1, pp. 312-362.

Shamsudin, A.Z. and Dillon, T.S. (1991) *NetManager: A Real-Time Expert System for Traffic Management*, Technical Report #15/91, Dept. of Computer Science and Computer Engineering, La Trobe University.

Simpson, P. (1990) *Artificial Neural Systems*, Pergamon Press, New York.

Stevenson, W.J. (1991) *The Use of Artificial Neural Nets in Mechanical Engineering*, Technical Report #AERO 91/193, Aeronautical Systems Technology, Council for Scientific and Industrial Research, South Africa.

Tzafestas, S. (ed.) (1993) *Expert Systems in Engineering Applications*, Springer Verlag, Berlin and Heidelberg.

Zahedi, F. (1990) A method of quantitative evaluation of expert systems, *European J. of Operational Research*, vol. 48, pp. 136-147.

Chapter 3   Statistical Process Control

Statistical Process Control (SPC) is the use of statistical problem solving measures to improve the quality of products or processes. The word **statistical** implies the collection, representation and interpretation of data using statistical methods such as control charts. **Process** refers to any sequence of operations which continuously utilises machines, methods, materials and manpower. **Control** implies taking care of the process by measuring and correcting its performance [Owen, 1989; Schilling, 1990]. This chapter gives an overview of SPC and control charting. The chapter describes XPC, an expert system for statistical process control. Finally, it discusses the selection of control chart types and reviews the use of intelligent advisor programs for control chart selection.

3.1 Statistical Process Control (SPC) and Control Charting

SPC is mainly employed in on-line production. Although it uses techniques developed in the 1920s as part of **statistical quality control**, it brings a new dimension of defining the process, controlling it and improving it. Improvement of the process is based on preventing problems before they occur rather than waiting for them to happen and finding the solution afterwards. This necessitates expert knowledge about the process and skills to implement SPC. That is why SPC requires the process operators, supervisors and managers all to be involved in controlling the process [Owen, 1989; Ford Motor Company, 1984].

The need for SPC expertise also arises from the manufacturing process itself. A good production system requires equipment that is capable of meeting precision and accuracy requirements (derived from the customer specifications), qualified operators, efficient working methods and proper materials. Even the best designed and managed production systems cannot manufacture every part or provide all services uniformly and consistently. Variability is an unavoidable element of manufacturing. Any two finished products or product characteristics are seldom identical [Evans and Lindsay, 1989; Grant and Leavenworth, 1988]. The difference may be large or immeasurably small. This variability may be the result of either internal or external factors.

Careful analysis of the process and manufacturing system will define the nature of the variation. The process is said to be in-control if the variation is within the naturally unavoidable and expected (inherent) variation level. The causes contributing to this type of variation are known as *common* or *random causes*. If the variation of the process is higher than the inherent variation level, it is due to external factors such as miscalibrated equipment, tool wear, shift to shift differences among the operators, wrong machine settings or inconsistent material compositions. These causes are called *special* or *assignable causes*. It should be noted that assignable causes can be controlled and corrected by the process operator or supervisor through careful analysis and examination of the process. The correction should yield a reduction in process variations and improve the process and product quality.

SPC provides methods of measuring product and process variations by means of control charts which, through a graphical representation of the process variables, are the basic tools for monitoring process quality characteristics and eliminating problems of not meeting the required quality specifications. The main aim is to keep the process within the natural variation limits through performing detection, diagnosis, correction and prediction. However, SPC is under-utilised in many industries because of the difficulties for operators to interpret control charts and identify assignable causes for out-of-control situations. This has led to an interest in systems which can capture and combine expert knowledge to facilitate the successful use of SPC.

Control charts show the mean value and the upper and lower allowed variation limits (natural variation limits) of the quality characteristic under consideration (see Figure 3.1). These limits, usually taken as the mean value plus and minus three standard deviations, represent the boundaries of the range for unavoidable (inherent) variations. Three standard deviations are used because there is a high probability (99.73%) that a sample measurement will fall within this range if the process is in-control (the quality characteristic is assumed normally distributed based on the central limit theorem). Careful construction and analysis of the charts will determine the inherent variation of a process which is in-control and capable of making products to meet customer specifications. Problems which are related to control chart construction and interpretation are discussed in Dale and Shaw [1990].

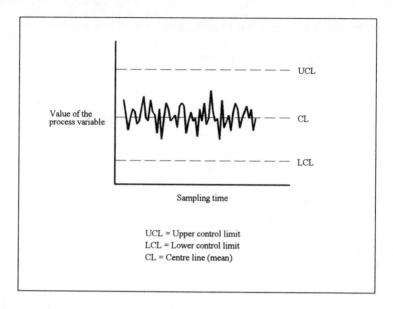

Figure 3.1 An example of a control chart

After the charts have been constructed, they are employed for on-line process monitoring. The most popular control charts are the mean and range charts. The mean chart is used to monitor the mean of a selected quality characteristic sampled from the process. The range chart is employed to control the dispersion or range of the samples. After plotting the on-line process data on the charts, the process supervisor determines whether the process exhibits any abnormal variation. If so, the next step is to find and correct the problems causing the abnormal variation in order to bring the process statistically under control.

The quality control literature has provided a set of control rules to interpret and detect out-of-control situations [Nelson, 1984]. For instance, since 99.73% of the means and ranges should be within the control limits (assuming a normal distribution of variables), when an on-going process produces a value outside these limits then there are two possible explanations for the deviation. First, the process is truly under control and this particular value belongs to the 0.27% of extreme values or, second, the process is out-of-control. The latter is the more likely conclusion. Thus, an out-of-control situation must be signalled. The corresponding control rule is:-

> IF any value is outside the control limits
> THEN the process is out-of-control.

Four of the other commonly adopted control rules are:-

      IF 6 consecutive values are running up or down
          THEN the process is out-of-control;

      IF 7 consecutive values are on the same side of the centre line
          THEN the process is out-of-control;

      IF 8 consecutive values are outside the 1 standard deviation limits
          THEN the process is out-of-control;

      IF two-thirds of the plotted values are in the middle third of the chart
          THEN the process is out-of-control.

When an out-of-control signal is issued after either inspecting control chart patterns or applying control rules, an investigation of the process is needed to make an accurate diagnosis of the problems that could have occurred.

There are several software packages developed to assist SPC coordinators. It has been noted that developments in computer technology have made it possible to devise computer-aided quality control systems, especially for SPC [Pham and Oztemel, 1992a and 1992b]. However, the majority of computer programs developed for SPC are off-line programs and are not designed to exercise process knowledge or provide interpretations of the process states. These functions are left to the quality experts with in-depth knowledge of the process. Since interpretation has to be done off-line by skilled quality controllers, immediate preventive action cannot be taken and faulty products will result. Sanders *et al.* [1989] have also highlighted the weaknesses of existing SPC software packages. They again state that the main problem with these packages is that their outputs can only be interpreted by very experienced operators or quality control engineers. This clearly implies the requirement for SPC specialists in manufacturing systems. They emphasise the need to employ knowledge-based systems or expert systems as an alternative way to overcome many of the difficulties in quality control. This has also been realised by other researchers [Iwata, 1988; Rao and Lingaraj, 1988; Spur and Specht, 1992]. As will be seen in this chapter, an expert system would be able to perform interpretation of quality data quickly. This would help to reduce scrap and production errors and lower training costs. Other benefits will be explored in the next section.

There have been a few applications of expert systems for SPC. Most of the work in this area is related to the selection of control chart parameters [Alexander and Jagannathan, 1986; Dagli and Stacey, 1988; Hosni and Elshennawy, 1988]. The systems developed suggest suitable control chart procedures for the problem presented to them. There are other intelligent systems which concentrate on different facilities of SPC involving, for example, cause and effect analysis or automatic interpretation of the control charts to detect out-of-control situations. These systems are mainly designed for off-line use to assist quality specialists rather than process operators and do not even incorporate process specific knowledge. For example, Bourne *et al.* [1989] have shown how cause and effect diagrams can be mapped into a belief network addressing several conceptual ideas. Evans and Lindsay [1988], Cesarone [1991] and Lall and Stanislao [1992] have developed expert systems to interpret control charts automatically. All of these systems are designed to be used off-line. One of the most powerful systems in this field has been developed by Dagli and Smith [1991]. It is capable of selecting suitable control charts and automatically interpreting them. However, it is also designed to be used off-line. An example of a completely automated on-line intelligent statistical process control system is described in the next section.

3.2 XPC: An On-line Expert System for Statistical Process Control

Expert system technology has been adopted to develop a software tool for Statistical Process Control which does not suffer from the previously mentioned deficiencies of existing SPC software. This section describes this expert Statistical Process Control tool which has been appropriately named XPC. The main functions of XPC are to perform control charting activities for $\overline{X}$ (mean) and R (range) charts and diagnose situations when the process goes out-of-control. The program is able to carry out on-line Statistical Process Control and troubleshooting by incorporating specific process knowledge as well as standard SPC knowledge. In the following, the structure of XPC is first introduced and the functional components are then explained in detail. Finally, examples related to a test application are presented.

A commercially available expert system shell, Leonardo [Bezant, 1992], has been used to develop the system. Leonardo is a mid-size hybrid tool with an inference engine implementing both forward and backward chaining. It represents knowledge using rules, objects, frames, classes and procedures. The facilities provided also include uncertainty management using Bayes' rule and certainty factors, built-in mathematical, statistical and graphical packages, interfaces to conventional programming languages (C, Fortran, Pascal), a spreadsheet package (Lotus 1-2-3) and database management systems (dBase III, dBase IV).

3.2.1 Structure of XPC

Figure 3.2 shows the structural components of XPC. They are its knowledge base, inference engine, user interface and statistical package.

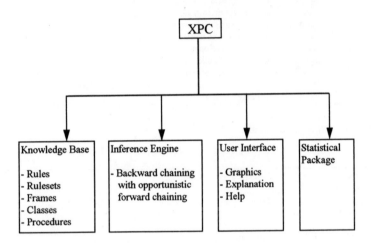

Figure 3.2 Overall structural components of XPC

The Knowledge Base: This is the portion of XPC that embodies general knowledge regarding statistical process control charting as well as specific knowledge about the process to be controlled. In total, the knowledge base comprises a main ruleset, 230 rules, 20 sub-rulesets, 340 object frames, a class and 29 procedures. These knowledge elements are defined below.

Main ruleset and sub-rulesets: The main ruleset, the basic component of the knowledge base, comprises a list of rules. When the number of rules is high, as it is in this case, the main ruleset is divided into sub-rulesets to facilitate knowledge base management and increase run-time efficiency. A sub-ruleset assembles the rules relevant to a particular situation and is invoked only when that situation exists. Rules perform a variety of functions. They can "assign" values to objects which can then be used by other rules, "execute" a procedure, "invoke" an input/output screen, "create" messages and "ask" the user for values of variables etc. Several antecedents and consequents can be incorporated into each rule using the AND or OR conjunctions. Examples of a rule and a ruleset in XPC are given in Figures 3.3 and 3.4 respectively.

```
IF seven_up=1
        THEN       cause7 includes "Non conform. plan. maint. sched.";
                   cause7 includes "Temp. taken in different places";
                   cause7 includes "Different pyrometers are used";
                   cause7 includes "Pyro. recalib. without author.";
                   cause7 includes "Temp. reset. without authority";
                   cause7 includes "Heater malfunction";
                   cause7 includes "Faulty thermocouple";
                   cause7 includes "Changes in material compos.";
                   cause7 includes "Change of operator";
                   cause7 includes "Rise in ambient temp.";
                   cause7 includes "Untrained operator";
"seven up" is an object that takes the value 1 if 7 measurements are above the mean value.
"cause7" is a list object that stores the related causes.
```

Figure 3.3 An example of a rule

```
        Name              : Tolerance_determination
        Long Name         :
        Type              : Text
        RuleSet           :
        IF start is yes
                THEN ask input_unit;
                input_unit_definition is done
        IF input_unit_definition is done
        AND input_unit is data_file
                THEN run proc_readtol(record);
                tolerance_definition is done
        IF input_unit_definition is done
        AND input_unit is keyboard
                THEN ask upper_tolerance_limit;
                ask lower_tolerance_limit;
                run proc_writetol(record);
                tolerance_definition is done
        IF input_unit_definition is done
        AND input_unit is default
                THEN upper_tolerance_limit=100.00;
                lower_tolerance_limit=60.00;
                tolerance_definition is done
```

Figure 3.4 An example of a ruleset

Objects and object frames: Associated with a rule are "objects", which are standard data modules in Leonardo. Numerical variables, string variables and lists are all represented as objects. The kind of data stored in an object is identified by the type of operator it is used with. For example, the "is" operator defines a text object, the "=" operator, a numerical object and the "include" operator, a list object. Values are assigned to an object using these and other similar operators.

All of the information relating to an object is stored in a "frame". This is a special knowledge structure for describing one or more values of the attributes of an object. An object frame is composed of "slots" in which the information is stored. There are 3 types of slot: protected slots which cannot be modified (e.g. the slot containing the name of the object), default slots which are automatically generated when an object is created (e.g. the "Query Prompt" slot for storing questions which may be posed to the user) and optional slots which can be added by the user when required (e.g. the "Preserve" slot for preserving the value of the object from a previous run). Examples of an object frame and the effect of activating it are given in Figures 3.5 and 3.6 respectively.

Name	: Job_type
Long Name	:
Type	: Text
Value	:
DerivedFrom	:
DefaultValue	: Process
ForbidUnk	:
AllowedValue	: Construction, Process, Update
Preserve	:
QueryPrompt	: What is the job to be performed?
QueryPreface	: Please select the appropriate job to be performed by highlighting your choice using the arrow keys. The default job type is "process" for on-line control
Expansion	: This choice allows you to design new charts. This choice allows you to monitor the process on-line. This choice allows you to update existing charts.
Commentary	:
Introduction	:
Conclusion	:

Figure 3.5 The frame of object "Job_type"

Please select the appropriate job to be performed by highlighting your choice using the arrow keys. The default job type is "process" for on-line control.		
What is the job to be performed?		
Construction		
Process	====>	This choice allows you to monitor the process on-line
Update		

Figure 3.6 Elaboration of the frame of object "Job_type"

Classes: If an object has to share information with other objects or if there is a kind of property inheritance between them, then a class object can be used. Objects from the same family sharing similar information are identified as the members of a class object. Each class has a "Members" slot where the members of the class are listed and a "MemberSlots" slot where common features of the members are identified. Related feature values for each member object are stored under the "MemberSlots" slot of that particular member. A class object and its frames are used to reduce the number of rules needed as the same rule for the class object applies to all of its members. Examples of a class object and a member object are shown in Figures 3.7 and 3.8 respectively.

Name	: Faults
Long Name	:
Type	: Class
Members	: Mislocated_shell, Undercure, Excessive_flash, Cold_runner_separation, Inclusion, Bond_failure, Air_traps_and_blisters, Omitted_shell
MemberSlots	:
Place	:
Fault_type	:
Symptom	:

Figure 3.7 An example of a class object

Name	: Mislocated_shell
Long Name	:
Type	:
IsA	: Faults
MemberSlots	:
Place	: Mould
Fault_type	: Defective_mouldings
Symptom	: Damaged_mould

Figure 3.8 An example of a member object

Procedures: Procedures are sets of instructions performing routine algorithmic tasks. They are coded in a high-level procedural language similar to Pascal and are able to perform complex arithmetic computations. They are also used to control the display screen and to interface with external programs and databases. A procedure can be called from a rule, a frame or another procedure. A procedure frame has different parameter slots: read-only parameter slots ("Accepts" slots), read-write parameter slots ("Returns" slots) and local parameter slots ("Local" slots). Read-only slots are locations for defining the parameters passed to the procedure. The values of these parameters are not updated by the execution of the procedure. Read-write slots allocate space for the parameters passed to the procedure. Their values can be updated during the running of the procedure. Local slots are used to define objects which are local only to the procedure employed. Additionally, there is an "External" slot which stores the name of an external program to be called and executed. The procedural language of Leonardo provides many built-in functions such as string handling statements, conditional statements and screen handling statements. An example of a procedure is given in Figure 3.9.

Inference Engine: The method of inferencing adopted in XPC is a combination of the backward and forward inferencing methods called "backward chaining with opportunistic forward chaining" [Bezant, 1992]. With this method, backward chaining is first performed to instantiate the "objects" in a rule or a user query and then forward chaining is carried out to maximise the use of this data.

User Interface: The graphically-based user interface handles communication between the user and the expert system. Quality control information is interactively obtained from the user. Help facilities and explanation facilities are also provided. Both control charts (mean and range charts) are displayed on the screen of the XPC computer on a continuous basis and updated dynamically after every measurement. This means that standard deviations and process capability indices are computed on-line and presented to the user.

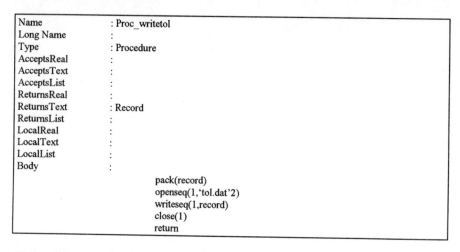

Name	: Proc_writetol
Long Name	:
Type	: Procedure
AcceptsReal	:
AcceptsText	:
AcceptsList	:
ReturnsReal	:
ReturnsText	: Record
ReturnsList	:
LocalReal	:
LocalText	:
LocalList	:
Body	:

```
pack(record)
openseq(1,'tol.dat'2)
writeseq(1,record)
close(1)
return
```

Figure 3.9 An example of a procedure

Statistical Package: This allows the system to perform standard statistical computations such as the derivation of means, standard deviations, averages, variances and minimum and maximum values.

3.2.2 Functions of XPC

The functional components of the system are shown in Figure 3.10. The system consists of four modules. The first module is a construction module which ascertains process parameters (such as tolerances, sample size and applicable process control rules) and constructs the charts to determine inherent variation levels. The second module is a capability analysis module that performs process capability analysis to ensure that the control charts constructed in the first module are compatible with customer specifications. This module is automatically invoked when the construction or modification module is activated. The third module is an on-line interpretation and diagnosis module. It monitors the process during production and locates quality problems. The fourth module is a modification module which updates the chart parameters to maintain control over the process.

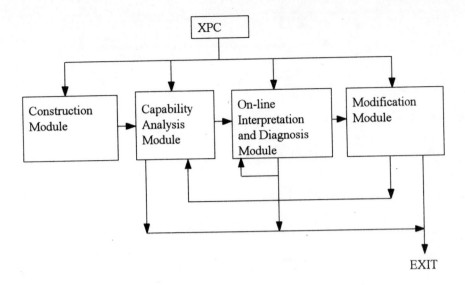

Figure 3.10 Functional components of XPC

Construction Module: The method of constructing the range (R) and mean ($\bar{X}$) control charts is summarised in Figure 3.11. An initial set of N samples (N = 25; sample size = 5) is obtained from the manufacturing process being monitored. The statistics package of the system is used to compute the means ($\bar{X}$), ranges (R) and the overall averages μ and $\bar{R}$. Trial limits for the charts are then set as follows:-

$$UCL_R = \bar{R} + 3\sigma_R \tag{3.1}$$

$$LCL_R = \bar{R} - 3\sigma_R \tag{3.2}$$

$$UCL_{\bar{X}} = \mu + 3\sigma_{\bar{X}} \tag{3.3}$$

$$LCL_{\bar{X}} = \mu - 3\sigma_{\bar{X}} \tag{3.4}$$

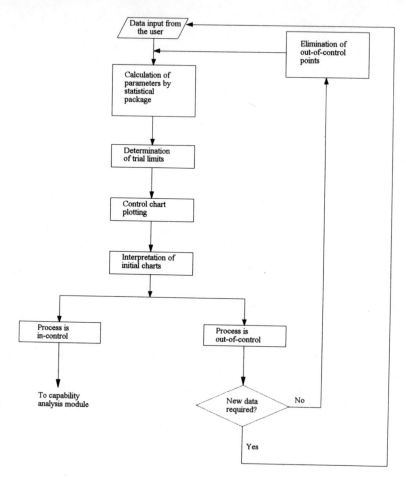

Figure 3.11 Overall structure of the construction module

where:-

$$\bar{R} = \frac{1}{N} \sum_{i=1}^{N} R_i \qquad (3.5)$$

$$\sigma_R = \sqrt{\frac{\sum_{i}^{N} (R_i - \bar{R})^2}{N - 1}} \qquad (3.6)$$

$$\mu = \frac{1}{N} \sum_{i=1}^{N} \overline{X}_i \qquad (3.7)$$

$$\sigma_{\overline{X}} = \sqrt{\frac{\sum_{i}^{N} (\overline{X}_i - \mu)^2}{N-1}} \qquad (3.8)$$

UCL_R, LCL_R, $\text{UCL}_{\overline{X}}$ and $\text{LCL}_{\overline{X}}$ are trial values for the control limits on the range and mean charts respectively.

The initial data set is plotted against these limits. The plots are automatically interpreted by XPC to determine if the process is in-control. If there are one or more out-of-control situations then new limits are calculated by either using a new data set from the process or eliminating those points that cause out-of-control situations from the existing data set. This procedure is repeated until both charts are constructed for a process which is in a state of statistical control. Note that the range chart is constructed before the mean chart as the latter cannot be used unless the range of the process characteristic of interest is in-control.

Capability Analysis Module: Capability analysis is performed only when the process is in a state of statistical control. This analysis, which is described in Figure 3.12, is carried out to ensure that the manufacturing process is able to meet the customer requirements. This is done by analysing the relationship between the distribution of the process data and the tolerance limits. Figure 3.13 shows different types of relationships that could occur. It is clear from this figure that, even if the process seems under control, it may not be able to meet customer specifications. The relationship of interest is formulated in terms of capability indices, namely:-

$$C_p = \frac{UTL - LTL}{6\sigma} \qquad (3.9)$$

$$C_{pu} = \frac{UTL - \mu}{3\sigma} \qquad (3.10)$$

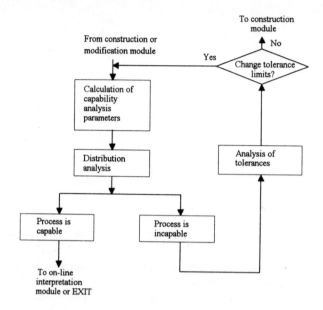

Figure 3.12 Overall structure of the capability analysis module

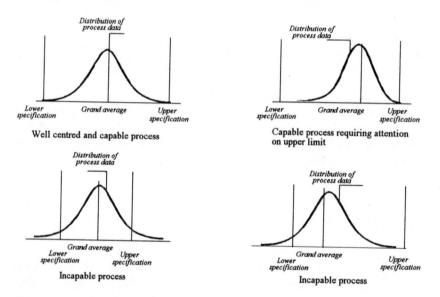

Figure 3.13 Examples of relationships between process data distributions and tolerance (specification) limits

$$C_{pl} = \frac{\mu - LTL}{3\sigma} \tag{3.11}$$

$$C_{pk} = \min(C_{pu}, C_{pl}) \tag{3.12}$$

where C_p is the overall capability index of the process, C_{pu} and C_{pl} are the capability indices associated with the position of the distribution of the process relative to the tolerance limits, C_{pk} is the worst case capability index and *UTL* and *LTL* are the upper and lower tolerance limits specified by the customer.

Capability indices measure the extent to which a process is able to meet customer specifications. Their values should be greater than or at least equal to 1. An index equal to 1 or higher means that the process is capable of performing within the specification limits. As the process variability decreases, the corresponding value of the capability index increases. An index of 1.33 is generally adopted as the target. Some examples of the rules for interpreting these indices are given in Figure 3.14. During the capability analysis, if it is found that the process is incapable then analysis of the specified tolerance limits will determine whether to reactivate the capability module or return to the construction module (see Figure 3.12).

IF $C_p > 1$ AND $C_{pk} < 1$
   THEN the process is capable of meeting the specification limits but the process mean needs to be moved as its value is not correct

IF $C_p > 1$ AND $C_{pk} > 1$
   THEN the process is capable of meeting customer specification limits. Some variability could still be allowed

Detailed explanation regarding individual specification limits (upper limit and lower limit) can be provided by analysing the C_{pl} and C_{pu} indices.

Figure 3.14 Examples of rules for capability analysis

On-line Interpretation and Diagnosis Module: On-line control is based on the following reasoning. If the process is in-control then 99.73% of all quality characteristic measurements will fall within the natural control limits. Thus, as previously stated, when an ongoing process produces a measurement outside these limits there could be two possible explanations. First, the process is truly under control and that particular measurement belongs to the 0.27% of extreme values. Second, the process is out-of-control. The more likely conclusion is the latter and therefore the system should immediately produce an out-of-control alarm signal. On-line monitoring in this way can be used by the operator to take immediate preventative action (usually performed off-line) to maintain product quality before components deviate from their acceptable level of quality. Figure 3.15 depicts the structure of the on-line interpretation and diagnosis module.

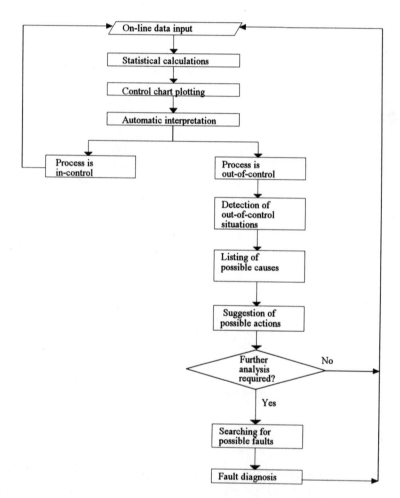

Figure 3.15 Overall structure of the on-line interpretation and diagnosis module

The function of the interpretation part is to determine whether the process is out-of-control or behaving correctly, based on measurements of the quality characteristic. This part accepts on-line process signals and applies the control rule set to interpret them for any out-of-control situation. If the process is out-of-control, an immediate warning signal is given to the operator and the knowledge base is searched for the likely causes of that situation and possible actions to correct it. Examples of an out-of-control situation and a list of possible causes of that situation are illustrated in Figures 3.16 and 3.17 respectively. A typical action table is given in Figure 3.18.

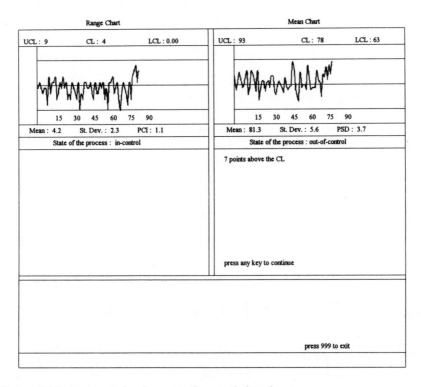

Figure 3.16 An example of an out-of-control situation

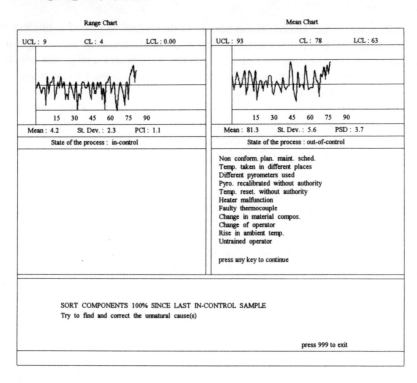

Figure 3.17 Possible causes of the out-of-control situation shown in Figure 3.16

	Capability index (C_p)		
	Poor < 1.00	Good 1.0 to 1.33	Excellent >1.33
Process is in-control	100% inspect	Accept parts	Accept parts
Process is out-of-control but all individuals in the sample are within the specifications	100% inspect	TRY TO CORRECT SPECIAL CAUSES Sort components 100% since last in-control operation	Accept parts
Process is out-of-control and one or more individuals in the sample are outside the specifications	100% inspect	TRY TO CORRECT SPECIAL CAUSES Sort components 100% since last in-control operation	

Figure 3.18 A typical action table

If further analysis is required by the operator, the system activates the diagnosis part. This is a Leonardo program containing knowledge about all possible process faults. Each fault is considered to be a member of a class object called "faults". Faults are described in terms of the place of occurrence (for example, on the product or the machine that produced it), type and symptoms. As a result of interaction with the operator, possible faults are determined and detailed actions are suggested to correct them. More than one out-of-control situation can be detected and several faults diagnosed at the same time. If an out-of-control situation that has been detected occurred by chance (with a probability of 0.27%) the system retains this information as a "point identifier". This information is used by the modification module.

Modification Module: Since the manufacturing process conditions upon which the original chart construction is based may be subject to change, the mean and the range of the process can vary during production. It is desirable to check and update the parameters of the charts to maintain control over the process and to improve product and process quality. Since process variability is reduced by corrective actions, the new parameters of the charts will also show an improvement in the capability of the process (by an increase in the capability indices).

Figure 3.19 illustrates the structure of the modification module. The first task that has to be performed by this module is to determine points representing truly out-of-control situations or aberrations which have previously been detected by analysing the point identifiers. Genuine out-of-control points have to be removed from the data set in order to define correct control limits because the latter would be wider if those points were included. Similarly, the control limits would be narrower if other points, which seem to be out-of-control but are not, are incorrectly eliminated. To modify the charts, new trial limits are calculated using the last 25 sample measurements of the process. These limits are then used against those 25 samples to interpret them.

If the process seems to be out-of-control with respect to the new limits then the trial limits are updated by eliminating out-of-control points. This procedure is repeated until the process is in-control. Afterwards, the modification module activates the capability analysis module to ensure that the distribution of the process is still compatible with the customer requirements.

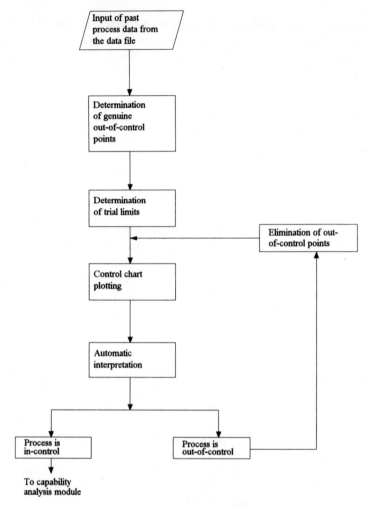

Figure 3.19 Overall structure of the modification module

3.2.3 Test Application and Discussion

In the first test application of XPC, the mould temperature in a polymer injection moulding machine was controlled. Expertise in injection moulding was acquired from a local company and embodied into the knowledge base of the expert system. General SPC knowledge was derived from the quality control literature [Evans and Lindsay, 1989; Grant and Leavenworth, 1988]. Specific process knowledge was provided by two foremen, two engineers and a quality specialist of the company where the test application of XPC was implemented. This knowledge included various out-of-control situations, possible causes for them, recommended actions and process faults to be diagnosed.

Control charting had previously been performed manually in the subject company. The operator was provided with blank mean and range charts which could accommodate 20-25 data points. The temperature was read every 4 hours and plotted on the charts. After the charts had been filled in, which required approximately one week, they were collected by the quality control engineer and interpreted for any out-of-control situations. The parameters of the charts were updated with this new information and blank charts were given to the operator again to be filled. The operator informed the quality engineer if the process had produced any temperature readings outside the limits. However, if there were other types of out-of-control situations, they had to wait to be determined by the quality engineer.

XPC has solved this problem of delays by performing real-time on-line automatic interpretation of the charts as well as providing diagnostic information. It can inform the operator of what actions to take as soon as the process goes out-of-control thus ensuring that the process quickly returns to a state of statistical control. The main benefit obtained is the capability of giving real-time advice to the operator in addition to reducing decision making delays. XPC is also a good tool for training the process operator.

In this particular process, other benefits that have resulted from XPC are:-

- the skill level requirements of the operator have been reduced;
- operational speed has been increased;
- human errors in computation and control charting have been eliminated;
- process variation has been decreased;
- the target overall process capability has been achieved;
- product quality has been improved;
- faulty products and scrap rates have been reduced.

3.3 Intelligent Advisors for Control Chart Selection

As previously explained, control charts are effective tools for monitoring the capability and the state of a process. There are more than 14 different types of control charts with variations depending on the application. Some of them are used very often while others are very specialised and rarely employed. Each control chart provides different types of information. The well-known charts are listed below:-

- Mean ($\overline{X}$) chart;
- Range (R) chart;
- s-chart;
- σ-chart;
- Individual mean chart;
- Moving ranges chart;
- Median chart;
- Mid-range chart;
- CUSUM (cumulative sum) chart;
- EWMA (exponentially weighted moving average) chart;
- Geometric moving average chart;
- p-chart;
- np-chart;
- c-chart;
- u-chart.

There are two basic kinds of data which may be taken into account in monitoring a process: (i) variable data which is derived from physical measurements (such as temperature, diameter, speed and time) and (ii) attribute data which includes information about the characteristics of a product (such as defective / non-defective and acceptable / unacceptable). Although control charts could be differentiated based on the data they refer to, deciding which control chart type is appropriate for a given process requires expertise as several charts could be applicable to a specific process. There are various factors which affect the selection procedure. They include:-

- sample size;
- resources available for control chart computations;
- type of information required by the chart;
- information on the number of defective items produced;
- methods of calculations;
- features of the quality characteristic to be monitored;
- lot size;
- sampling interval.

The following are two examples of general rules showing how this information is used when making decisions about the type of the charts.

IF data available is 'attribute'
AND sample type is 'percent'
AND sample size is 'averageable' OR 'constant'
      THEN suggested chart is 'p-chart'

IF measurements are 'possible (variable data available)'
AND time required to obtain measurement is 'small'
AND detection of shift in the mean and variance is 'required'
AND high sensitivity to process shifts is 'not critical'
      THEN suggested charts are 'Mean and Range charts'

Sometimes, an optimisation routine is attached to a selection system to determine optimum chart parameters such as sample size, upper and lower control limits and sampling interval. This is only possible when the information required by the optimisation routine is available. This information, which may not always be easy to provide, includes cost of sampling (fixed or variable), cost of false alarm, cost of investigating an out-of-control situation, penalty cost associated with production when the process is out-of-control, number of out-of-control causes for a certain period of time, probability of a false alarm, sampling time interval, time required to find out an assignable cause or the expected duration of an out-of-control situation and time required to take a sample.

An example of an intelligent control chart advisor is ASCC (Advisor for Selection of Control Charts) developed by Dagli and Stacey [1988]. ASCC advises the user to choose a control chart based on his answers to a series of questions. ASCC also utilises an optimisation routine to define the parameters of the charts. The selection process is mainly based on chart type (data type, attribute or variable), sub-group size, sampling type, sample size and calculation ability (type of calculation method, computer, hand calculations etc.). The system can recommend more than one control chart with related confidence factors. Similar systems have been developed by others including Dagli and Smith [1991], Lall and Stanislao [1992], Alexander and Jagannathan [1986] and Hosni and Elshennawy [1988]. The general selection procedure of Hosni and Elshennawy is given in Figure 3.20.

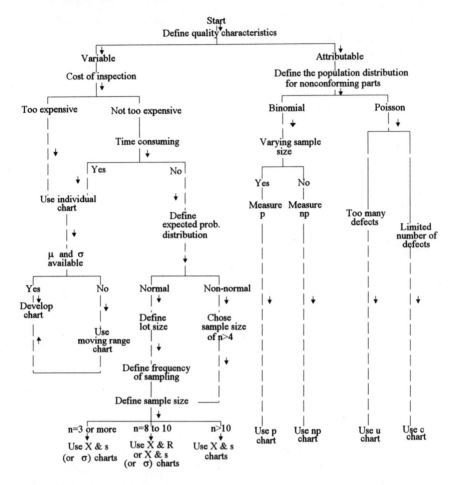

Figure 3.20 Procedure suggested for selecting a control chart [Hosni and Elshennawy, 1988]

3.4 Summary

This chapter has provided an overview of Statistical Process Control with respect to control charting. It has pointed out that the interpretation of quality control charts, diagnosis of out-of-control situations and determination of appropriate remedial actions require expert knowledge. An expert system called XPC, which has been developed to embody expert process knowledge for statistical process control, has been presented. The structural components of the system such as the knowledge base, inference engine and user interface have been explained. Functional details of the system, including its chart construction, on-line automatic interpretation and diagnosis and modification modules have been described. Qualitative results for a test application on an injection moulding process have been presented. The selection of suitable control chart types has been discussed and the use of intelligent advisors for control chart selection briefly reviewed.

References

Alexander, S.M. and Jagannathan, V. (1986) Advisory system for control chart selection, *Computers and Industrial Engineering*, 10(3), 171-177.

Bezant Ltd. (1992) **LEONARDO Reference Manual**, Bezant Ltd., Wallingford, Oxon.

Bourne, J., Liu, H., Orogo, C. and Uckun, S. (1989) Intelligent systems for quality control, *Proc. 3rd Int. Conf. on Expert Systems and the Leading Edge in Production and Operations Management*, Head Island, South Carolina, 321-332.

Cesarone, J. (1991) QEX: An in-process quality control expert system, *Robotics and Computer Integrated Manufacturing*, 8(4), 257-264.

Dagli, C.H. and Stacey, R. (1988) A prototype expert system for selecting control charts, *Int. Journal of Production Research*, 26(5), 987-996.

Dagli, C.H. and Smith, A.E. (1991) A prototype quality control expert system integrated with an optimisation module, *Proc. of the World Congress on Expert Systems*, Orlando, Florida, 1959-1966.

Dale, B.G. and Shaw, P. (1990) Some problems encountered in the construction and interpretation of control charts, *Quality and Reliability Engineering International*, 6, 7-12.

Evans, J.R. and Lindsay, W.M. (1988) A framework for expert system development in statistical quality control, *Computers and Industrial Engineering*, 14(3), 335-343.

Evans, J.R. and Lindsay, W.M. (1989) *The Management and Control of Quality*, West Publishing Company, St Paul, MN.

Ford Motor Company (1984) *Statistical Process Control: Instruction Guide*, Ford.

Grant, E.L. and Leavenworth, R.S. (1988) *Statistical Quality Control*, McGraw Hill, New York, 6th edition.

Hosni, Y.A. and Elshennawy, A.K. (1988) Quality control and inspection - Knowledge-based quality control system, *Computers and Industrial Engineering*, 15(1-4), 331-337.

Iwata, K. (1988) Applications of expert systems to manufacturing in Japan, *Int. Journal of Advanced Manufacturing Technology*, 3(3), 23-37.

Lall, V. and Stanislao, J. (1992) Applying a coupled expert system to quality control charts, *Computers and Industrial Engineering*, 23(1-4), 401-404.

Nelson, L.S. (1984) The Shewhart control charts - Tests for special causes, *Journal of Quality Technology*, 16(4), 237-239.

Owen, M. (1989) *SPC and Continuous Improvement*, IFS Publications, Springer Verlag, Exeter, UK.

Pham, D.T. and Oztemel, E. (1992a) TEMPEX: An expert system for temperature control in an injection moulding process, *Quality and Reliability Engineering International*, 8, 9-15.

Pham, D.T. and Oztemel, E. (1992b), XPC: An on-line expert system for statistical process control, *Int. Journal of Production Research*, 30(12), 2857-2873.

Rao, H.R. and Lingaraj, B.P. (1988) Expert systems in production and operations management: Classification and prospects, *Interfaces*, 18(6), 80-91.

Sanders, B.E., Sanders, S.A.C. and Cherrington, J.E. (1989) Knowledge acquisition and knowledge-based systems as an aid to product quality control, *Advances in Manufacturing Technology IV, 5th National Conf. on Production Research*, Huddersfield, 364-368.

Schilling, E.G. (1990) Elements of process control, *Quality Engineering*, 2(2), 121-135.

Spur, G. and Specht, D. (1992) Knowledge engineering in manufacturing, *Robotics and Computer Integrated Manufacturing*, 9(4/5), 303-309.

Chapter 4 Control Chart Pattern Recognition

As discussed in the previous chapter, control rules are used to detect out-of-control situations by considering the very recent history of a process. However, to avoid such situations, it is necessary to monitor the long-term history as recorded in control charts. Patterns of variations in a control chart can reveal impending out-of-control situations and help to form cause-effect relationships to predict possible abnormalities in a manufacturing process. In this chapter, automatic control chart pattern recognisers utilising heuristic rules and neural networks as well as combinations of these techniques are described. Experimental results show that these systems are capable of identifying patterns and providing early detections of abnormal conditions with a high degree of accuracy.

4.1 Control Chart Patterns

A problem with control rules is that they use only the most recent process data thus ignoring any information contained in the history of the process. Analysis of overall patterns in a control chart overcomes this problem as these patterns reflect the longer term process history.

A control chart pattern can be classified as (i) a *natural pattern* or (ii) an *unnatural pattern*. A natural pattern (usually called a normal pattern) is a random array of data points that possess several characteristics based on the underlying distribution of the data over a time sequence. Normal patterns usually have the following characteristics [Wadsworth *et al.*, 1986]:-

- Most of the data points occur near the centre line;
- A few points occur near the control limits;
- None (or few) of the points exceed the control limits;

The points occur in a random manner with no clusters, trends or other departures from a random distribution.

An unnatural pattern can be defined as a pattern which lacks the characteristics of random data points. The points in this type of pattern tend to fluctuate widely or fail to balance around the centre line. When an unnatural pattern occurs, it implies the existence of some assignable causes related to the process. It is important to know the method of observation and the characteristics of the process being observed to discover the causes of abnormalities. The main unnatural patterns that signify out-of-control situations include a jump or a change in the mean level, a trend or steady change in the mean of the process and recurrent cyclic patterns. A review of common control chart patterns is presented in most statistical quality control handbooks. For example, see [Western Electric Co., 1985].

4.2 A Knowledge-Based Control Chart Pattern Recognition System

Control chart pattern recognition using knowledge-based technology has been employed by a number of researchers. Many off-line recognition systems have been developed to classify patterns in control charts. They are mostly based on statistical hypotheses, heuristics and templates. This makes them rely on assumptions requiring prior process knowledge. Swift [1987] has described a knowledge-based control chart pattern recogniser. The system employs statistical hypotheses and is designed for off-line use. A drawback of the system is that it assumes an in-control state always follows an out-of-control state whereas, in practice, once a process has gone out-of-control, it is unlikely to return to an in-control state without corrective intervention. Similar systems have been reported for control chart pattern recognition using templates [Cheng, 1989; Cheng and Hubele, 1989] or control theory [Love and Simaan, 1989; Simaan and Love, 1990] instead of statistical hypotheses.

Pham and Oztemel [1992a] have described an on-line control chart pattern recogniser utilising heuristic rules and statistical hypothesis. The system can recognise a range of patterns including normal patterns, increasing and decreasing trends (ramps), upward as well as downward shifts (steps) and cyclic patterns (see Figure 4.1).

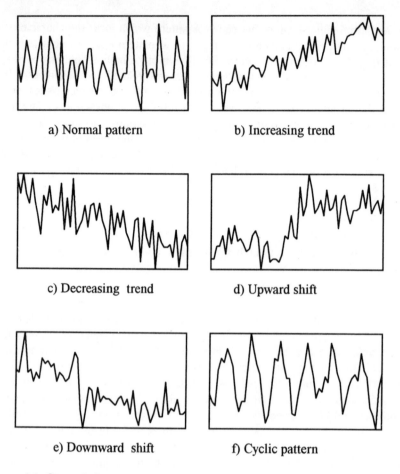

a) Normal pattern

b) Increasing trend

c) Decreasing trend

d) Upward shift

e) Downward shift

f) Cyclic pattern

Figure 4.1 Control chart patterns

The system is based on a number of assumptions regarding the patterns and the process under observation, such as:-

- the mean and standard deviation of the process are known;
- only one type of pattern is present in a given time window;
- the slope of trend-type patterns, the shifting position of shift-type patterns, and the period and amplitude of cyclic patterns are not known;
- the maximum allowed deviation from the process mean (process mean deviation threshold) for a normal pattern, the minimum average slope (slope threshold) for a trend pattern and the maximum least-mean-square linear regression error (error threshold) for trend, cyclic and shift patterns are known.

4.2.1 On-line Pattern Recognition and Classification

Each pattern to be classified is a time series consisting of 60 points. These are the 60 most recent average measurements of the quality characteristic being monitored. Continuous on-line updating of the time series is carried out by adding the latest available average and removing the oldest average at each sampling time. To reduce noise in the pattern, a moving average operation is applied to the time series before other statistical computations are made.

These statistical computations include:-

- calculation of the mean of the series and testing if it is significantly different from the process mean;
- least square fitting of a straight line to the series;
- determination of the linear regression error;
- determination of the slope of the fitted straight line;
- calculation of the auto-correlation coefficients.

The calculation of moving averages is based on defining a moving window and computing the average values of the means inside the window. The window is then moved forward one point towards the end of the time series. During the development of the system, different window sizes were experimented with as explained in section 4.2.2.

The least-square method is used to fit a straight line to a series of moving averages using the standard equation given by [Robinson, 1981]:-

$$Y(t) = a + bt \tag{4.1}$$

where:-

$$a = \overline{Y} - b\overline{t}$$

$$\overline{Y} = \frac{1}{N}\sum_{t}^{N} Y(t)$$

$$\overline{t} = \frac{1}{N}\sum_{t}^{N} t$$

$$b = \frac{\sum_t tY(t) - \overline{Y}\sum_t t}{\sum_t t^2 - \overline{t}\sum_t t}$$

In the above equations, N is the number of points in the time series, $Y(t)$ is the moving average value at time t, a is the Y-intercept and b is the slope of the fitted line. The calculation of a and b in this way minimises the squared deviation of the individual moving averages from the line.

Auto-correlation coefficients show the correlation between observations at points separated by different time intervals. The following simplified equation is used to determine the correlation coefficient r_k between observations which are k time intervals apart [Chatfield, 1989]:-

$$r_k = \frac{\sum_{t=1}^{N-k} (Y(t) - \overline{Y})(Y(t+k) - \overline{Y})}{\sum_{t=1}^{N} (Y(t) - \overline{Y})^2} \qquad (4.2)$$

Recognising Normal Patterns: The mean of a normal pattern should not be much different from that of the process. This difference is defined by the following two statistical hypotheses [Evans and Lindsay, 1989]:-

H_0 : The overall mean is equal to the process mean
H_1 : The overall mean is not equal to the process mean

These two hypotheses are tested using $Z_{\alpha/2}$ and Z_1 (standard unit normal) statistics. α is the probability of rejecting H_0 when, in fact, it is true. $Z_{\alpha/2}$ is the critical value which is tabulated for different values of α and can be found in standard quality control books [Evans and Lindsay, 1989; Grant and Leavenworth, 1988]. Z_1 is the test statistic which is calculated as:-

$$Z_1 = \frac{\overline{Y} - m}{s/\sqrt{n}} \qquad (4.3)$$

where:-

μ is the mean of the process;
$\overline{Y}$ is the mean of the pattern;
σ is the standard deviation of the process;
n is the sample size.

The mean of the data is said to be significantly different from that of the process at the $(1-\alpha) \times 100\%$ level of confidence if:-

$$Z_1 > Z_{\alpha/2}$$

or

$$Z_1 < -Z_{\alpha/2}$$

This means that hypothesis H_o is to be rejected and hypothesis H_1 accepted. Through experimentation with different values of α, it is found that $\alpha = 0.10$ provides the best discrimination. In addition, a normal pattern must fit a straight line with a slope below the slope threshold and a least square regression error below the error threshold. Thus, if the mean of the pattern is not significantly different from the process mean and both the slope of the fitted line and the related regression error are below the respective thresholds, the given pattern is classified as normal.

Recognising Trend-type Patterns: It has been mentioned that statistically significant differences between the mean of a pattern and the mean of a process are indicative of a trend or a shift. The converse of this statement is not always correct. For example, when the mean only starts shifting towards the end of a pattern, the mean of the entire pattern would not be statistically different from that of the process. The same applies when there is a trend with a small slope. The recognition system should be able to handle such situations.

The following heuristic has been adopted for identifying trend patterns. If the slope of the fitted line is above the slope threshold and the regression error is less than the error threshold then the trend pattern is confirmed. A positive slope indicates an increasing trend and a negative slope, a decreasing trend. This heuristic enables the special cases mentioned above to be dealt with effectively.

Recognising Shift-type Patterns: The shift position in a shift-type pattern could be either at the beginning or the end of the pattern or at some intermediate point within it. The following conditions apply to the first case:-

• the pattern mean is significantly different from the process mean;
• the slope of the straight line fitted to the pattern is below the slope threshold;
• the regression error is less than the error threshold.

For the second case where the shift position is at some intermediate point, the following conditions hold:-

- the slope of the fitted line exceeds the slope threshold;
- the regression error for the fitted line is higher than the error threshold;
- the slope of the line fitted to the part of the pattern after the shift position is below the slope threshold (the shift position is taken as the first point from which n consecutive moving averages are significantly different from the process mean where n is the moving average window).

Recognising Cycles: If the least-square-error straight line fitted to a given pattern has a slope below the slope threshold and the error is above the error threshold then the series is likely to exhibit cyclic behaviour. It has been suggested that auto-correlation is one of the most effective ways to recognise such a behaviour in a signal [Chatfield, 1989]. Auto-correlation coefficients are therefore calculated for the pattern. If the sum of these coefficients is nearly zero (i.e. the auto-correlogram for the pattern is cyclic), the pattern is confirmed as a cyclic pattern.

Note that patterns not fitting the above mentioned descriptions are regarded as *unclassifiable* by the system.

4.2.2 Performance of the Recognition System

Performance analysis has been carried out to define the accuracy of the system. The classification accuracy of the system is calculated using the following equation:-

$$\text{Accuracy}(\%) = \frac{\text{Number of patterns correctly classified}}{\text{Total number of patterns tested}} \times 100 \qquad (4.4)$$

The system has been tested on patterns generated by a process simulator. In total, 1998 known patterns are employed in the test (333 patterns of each type). Moving average window sizes of 3, 4 and 5 points are used. Slope thresholds equal to 0.10/sampling interval, 0.15/sampling interval and 0.20/sampling interval and least-square-error thresholds ranging from 300 to 700 are tried. The best results obtained for different moving average window sizes are given in Figure 4.2.

Moving average window size	3	4	5
Slope threshold	0.15	0.15	0.15
Error threshold	700	700	500
Performance (%)	92.87	94.81	93.95

Figure 4.2 Pattern recognition results

Thus, the overall best classification (94.81%) is achieved with a moving average window size of 4, a slope threshold of 0.15 and an error threshold of 500. The number of patterns correctly classified is 1895 and the remainder are either incorrectly classified or unrecognised. These results show that the system is capable of recognising control chart patterns with a high degree of accuracy.

4.3 Using Neural Networks to Recognise Control Chart Patterns

The pattern recognition and classification capabilities of neural networks have been shown to be better than those of conventional techniques [Lippmann, 1989]. As mentioned earlier, the latter are mostly based on statistical hypotheses or templates. This makes the development process difficult because usually an expert is required who has prior process knowledge. Also, when arbitrary patterns are observed, the recognition process becomes more complex and the level of the knowledge needed increases. Neural-network-based pattern recognisers, on the other hand, perform identification and classification with minimum process knowledge requiring only examples of how different patterns are classified. Such pattern recognisers are able to generalise from the given examples. This enables arbitrary patterns to be readily classified. However, the problem with these pattern recognisers is that it is not easy to discover how and why they classify a pattern into a certain class. This is due to their implicit knowledge processing. Correct selection of the topology of the networks is also very important as there are no theoretical guidelines about the most appropriate network structure for a given task.

Using neural networks to recognise control chart patterns has not yet attracted much attention. Hwarng and Hubele [1991] have proposed a neural-network-based recogniser for classifying control chart patterns. They divide a control chart into 7 zones. The proposed system accepts only binary inputs and therefore the process data is pre-processed and encoded into binary form. Each input value is represented by a code of 7 digits. A digit is equal to 1 if the input value belongs to the zone that the digit represents and 0 otherwise. This system considers 8 process values at a time (making 56 inputs for the neural network). The same problem as with the use of control rules is encountered here again. Eight input values from the process are not sufficient to indicate its true behaviour. Increasing the number of measured values from the process will make the number of input units for the neural network increase by 7 times the number of added process measurements, which lengthens the training procedure. A different approach, developed by the authors of this book, is presented in this section. The proposed neural network accepts continuous data from the process. In this case, 60 process quality values are input to the network, instead of just 8. This larger number of process data enables a more accurate assessment of the state and behaviour of the process.

In this section, two types of neural-network-based pattern recognisers, for the classification of control chart patterns, will be introduced. The first type uses a Multi-Layer Perceptron (MLP) trained by supervised learning [Pham and Oztemel, 1992b; Pham and Liu, 1995]. The second type employs a Learning Vector Quantisation (LVQ) model which follows the reinforcement learning strategy [Pham and Oztemel, 1994]. The section also presents a modified LVQ procedure yielding a higher classification accuracy than achievable with existing LVQ algorithms. The pattern recognisers can evaluate data routinely collected for control charting to determine if an abnormal pattern exists. A commercially available neural network simulator, Professional II [NeuralWare, 1993], and the authors' own simulation algorithms, written in Microsoft C, are used to train and test the pattern recognisers.

4.3.1. Pattern Recognition using a MLP Model

Network Structure: The MLP pattern recogniser consists of three layers: an input layer, a hidden layer and an output layer. The input layer which receives the data (pattern) to be identified has 60 neurons, one for each pattern point. The hidden layer which extracts features from the input pattern comprises 35 neurons. This number is arrived at following experimentation with hidden layers of various sizes. The output layer which processes extracted features to obtain the pattern class has 6 neurons, one dedicated to each of the available classes (Figure 4.3). Other features of the network are detailed in Figure 4.4.

Pattern	Outputs					
	a	b	c	d	e	f
Normal (a)	1	0	0	0	0	0
Increasing trend (b)	0	1	0	0	0	0
Decreasing trend (c)	0	0	1	0	0	0
Upward shift (d)	0	0	0	1	0	0
Downward shift (e)	0	0	0	0	1	0
Cyclic (f)	0	0	0	0	0	1

Figure 4.3 Representation of the output categories

	Input layer	Hidden layer	Output layer
Number of neurons	60	35	6
Combination function	-	Sum	Sum
Activation function	-	Sigmoid	Sigmoid
Global error	-	Standard mean square error	Standard mean square error
Learning rule	-	Delta rule	Delta rule
Control strategy	-	Backpropagation	Backpropagation

Figure 4.4 Details of the MLP pattern recogniser

The neurons in the input layer have a unity activation function, which means that they simply transmit the scaled values of the pattern points directly to the hidden layer. The information processing in the hidden and output layers is implemented by a sigmoidal activation function. Inputs to the network are continuous and in the range 0-1. The network outputs are also continuous and in the same range. The outputs are thresholded so that values above a pre-set threshold (known as the selection threshold) are taken as 1 and those below the threshold are equated to 0. An output of 1 indicates that the pattern that produces it definitely belongs to the class represented by that output. Conversely, an output of 0 means that the pattern does not belong to that class.

Training data: Experimental evidence suggests that the generalisation capability of a network is strongly affected by the training data presented to it. A requirement for the training examples is that they should be consistent. 1500 patterns have been generated by the process simulator (250 of each pattern type) of which 498 are employed in the training set (83 of each pattern type). The remaining patterns are used to test the classification accuracy of the trained network. Both the training and test sets consist of pattern pairs, each made up of an input pattern and a corresponding output pattern indicating the class of the input pattern. Since in practice the parameters of the patterns are not known a priori, patterns are produced with randomly chosen parameters to train the network and test its generalisation ability. During the testing phase, the network is presented with new (previously unseen) patterns which are different from those used during the training phase.

Training Procedure: The network is trained using the Generalised Delta Rule as explained in Chapter 2; see also [Rumelhart *et al.*, 1986]. As previously mentioned, the knowledge of the network is stored in the connection weights between the neurons. The objective of training is to adjust these weights so that the application of a set of inputs produces the desired outputs. A number of issues regarding network training are considered:-

- the choice of learning parameters;
- the scaling of the inputs;
- the method of presenting the inputs to the networks;
- the method of initialising the connection weights.

Based on the different learning rates and momentum coefficients (the two main learning parameters) tried, low learning rates and high momentum coefficients have been found to be the most suitable combination. The values for the learning rate and the momentum coefficient are chosen as 0.3 and 0.8, respectively.

Each input pattern in the training set is scaled between 0 and 1 corresponding to the minimum and maximum input values. Scaling is useful for mapping the input values to the appropriate output levels. The following equation is used for scaling the input:-

$$\bar{y}(t) = \frac{y(t) - y_{min}}{y_{max} - y_{min}} \tag{4.5}$$

where:-

$\overline{y}(t)$ = scaled pattern value in the range 0 to 1;
$y(t)$ = pattern value at time t;
y_{min} = minimum allowed value (taken as 50);
y_{max} = maximum allowed value (taken as 110).

Usually, training input patterns are presented either sequentially or randomly to the network. With sequential presentation, patterns are shown to the network during training in the same order as they are stored in the training data file. When all the patterns have been presented once (i.e. when a training epoch has been completed), the next epoch is started and the patterns are presented to the network again in the same order. This procedure is repeated until the network has learnt all patterns in the training set sufficiently well. In random presentation, patterns are selected randomly from the training set and presented to the network until the learning is complete. Both of these methods have been experimented with and sequential presentation is found to facilitate training and is therefore adopted.

Multi-Layer Perceptron networks are sensitive to the initial values of the weights [Kollen and Pollack, 1990]. Also, the initial values should be very small for the network weights to converge. Therefore, all the weights are randomly initialised between -0.1 and 0.1. However, experiments have shown that the network could, in certain cases, also learn even with larger initial weights, for example between -2 and 2. The use of these larger initial weights makes the network more sensitive to the learning parameters and lengthens the learning time.

Test Results: The backpropagation algorithm is employed for training the network. After 200 training epochs the network can learn 100% of the patterns in the training set. The generalisation capability of the network, or the accuracy of classifying new patterns, is tested using a data set comprising 1002 previously unseen patterns.

At the end of every 50 training epochs, the network accuracy is measured using the test set and plotted for two different output selection thresholds values, namely, 0.8 and 0.9 (Figure 4.5). After 200 training epochs, the network can identify and classify 95.2% of the test set accurately, i.e. 954 out of 1002 unseen patterns are correctly recognised. This accuracy level remains unchanged beyond 200 training epochs for the selection threshold value of 0.8 and shows only a slight improvement for the threshold value of 0.9.

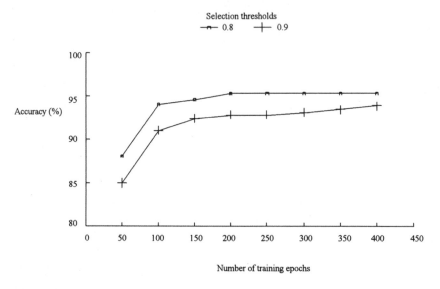

Figure 4.5 Performance of the MLP network

4.3.2 Recognising Control Chart Patterns using LVQ Networks

Network Structure: A number of LVQ network structures with different parameters and learning procedures have been experimented with. The networks finally adopted all have 60 input neurons (the input patterns consist of the 60 most recent mean values of the process variable to be controlled; therefore, one input neuron is dedicated to each mean value), 6 output neurons (one for each pattern category) and 36 Kohonen neurons (giving 6 reference vectors per category). A description of the network parameters is given in Figure 4.6.

At the beginning of training, the connection weights are set randomly between 0.1 and -0.1. Alternatively, in some cases, appropriate input patterns are assigned as the initial weights. The input vector components are scaled between 0 and 1. As in the case of the MLP network, the input vectors are presented sequentially to the network together with the corresponding output vectors identifying their categories.

Training Set: The same data sets as those used to train and test the MLP network are adopted for the various LVQ networks.

Number of Kohonen units for each output category	6
Initialisation range	-0.1 to 0.1
Scaling range	0 to 1
Learning range	0.05 to 0.01
Constant B	0.0001
Constant C	10

Figure 4.6 Details of the LVQ network parameters

Extended LVQ Training Procedure (LVQ-X):   The training procedure employed is largely based on the learning procedure originally developed by Kohonen [1990]. Note that in two of the existing LVQ models, only one weight vector is updated at each learning iteration (see Chapter 2). Even in the case of LVQ2 which updates two weight vectors at a time, it only does so under rare circumstances. In LVQ-X, the extended version of the LVQ learning procedure detailed below, two reference vectors are updated in most iterations resulting in a decrease in the learning time and an increase in the generalisation capability of the system.

The main idea of LVQ-X is to modify two candidate weight vectors. The first, called the "global winner", is the weight vector nearest to the input vector. The second, the "local winner", is the weight vector which is in the correct category and nearest to the input vector in that category. If the global winner is not in the correct category then it is pushed away from the input vector and at the same time the local winner is brought closer. This gives a chance for the correct neuron to win the competition in the next iteration. Obviously, if the global winner is also the local winner then only one weight vector needs to be updated. In this case, the weights are modified as follows:-

$$\mathbf{W_{new}} = \mathbf{W_{old}} + \lambda(\mathbf{X} - \mathbf{W_{old}}) \qquad (4.6)$$

where λ is the learning rate. If the global winner is different from the local winner then:-

$$\mathbf{W_{new}} = \mathbf{W_{old}} - \lambda(\mathbf{X} - \mathbf{W_{old}}) \text{ - for the global winner} \qquad (4.7)$$

and

$$\mathbf{W_{new}} = \mathbf{W_{old}} + \lambda(\mathbf{X} - \mathbf{W_{old}}) \text{ - for the local winner} \qquad (4.8)$$

The conscience mechanism, as described in Chapter 2, is also utilised in the proposed learning procedure. Both LVQ-X and the conscience mechanism are turned "off" towards the end of learning and LVQ2 is applied to refine the solutions.

Results: The generalisation ability of the network is tested using the same procedure as for the MLP network. The results obtained with the various LVQ networks are discussed below.

Standard LVQ: The initial results obtained with this network are poor because some neurons tend to win the competition too often. The initialisation procedure proves very critical. The network cannot learn the complete set of training patterns when the weights are randomly initialised. The accuracy levels achieved are all below 60%. Assigning some of the patterns in the training set as initial weight vectors increases the number of patterns that can be learnt. However, the network still does not achieve 100% learning. The classification accuracy levels of the network after 70 training epochs are 95.18% for the training data and 92.31% for the test data.

LVQ2: The initialisation procedure is again very critical. Patterns from the training set are assigned as the initial values of the weight vectors since the network cannot learn with random initial weight values. It is found that after only four training epochs the LVQ2 learning procedure is no longer applicable because the conditions set out in section 2.2.4 of Chapter 2 cease to hold and the weights in the network stop changing from that point onward. The overall accuracy levels of the network are 94.31% for the training set and 89.62% for the test set after 4 training epochs. The accuracy levels achieved when LVQ2 is applied to a network trained for 70 epochs using the standard LVQ algorithm are 96.18% for the training set and 92.61% for the test set. Again, the LVQ2 algorithm stops being applicable after four training epochs.

LVQ with a Conscience Mechanism: Applying a conscience mechanism to the standard LVQ model increases the learning capability of the network. It also decreases the dependence on using training examples as initial weights. With the same network structure as adopted for the other LVQ networks, a LVQ module with a "conscience" can learn 95.98% of the training set and correctly classify 92.71% of the test set following 70 training epochs.

LVQ-X: As mentioned above, none of the LVQ networks employed (standard LVQ, LVQ2 or LVQ with a conscience mechanism) can learn all the training patterns. The proposed LVQ-network, on the other hand, achieves 100% learning and a better classification accuracy in a shorter training time. Dependency on the initial values of the weight vectors is virtually eliminated and the performance of the network is almost the same for the cases of arbitrary initial weight values and initial weight values taken from the training set. At the end of 10 training epochs, the network can correctly classify 99.39% of the training set and 96.30% of the test set. After 20 training epochs, the overall recognition accuracy level increases to 100% for the training set and 97.70% for the test set. This clearly shows the superior performance of the modified version over the other LVQ procedures.

A comparison of the various LVQ models is given in Figure 4.7. The proposed LVQ-X network achieves even better classification accuracy than that obtained with the MLP network.

Pattern recogniser	Number of training epochs	Learning performance (%)	Test performance (%)
LVQ (Standard)	70	95.18	92.31
LVQ2	4	94.31	89.62
LVQ (Standard) + LVQ2	74	96.18	92.61
LVQ (with a conscience mechanism)	70	95.98	92.71
LVQ-X	20	100.0	97.70

Figure 4.7 Performance of various LVQ pattern recognisers

4.3.3 Discussion

Control chart patterns normally contain a random noise element. Therefore, it would be difficult to classify these patterns using simple heuristic rules with fixed and well-defined detection limits. Experiments have shown that the noise filtering and generalisation capabilities of neural networks make them suitable for this classification task. This is clearly demonstrated by the good identification results presented in this chapter. Both MLP and LVQ network models are able to classify control chart patterns with a high degree of reliability. Although the standard LVQ network has a relatively poor performance, after a slight modification, it achieves the best classification performance in a short training time. The modified procedure has enabled the network to perform classification with almost 98% accuracy.

4.4 Composite Systems for Recognising Control Chart Patterns

This section presents a number of composite pattern recognition systems with improved classification capabilities compared to the systems introduced in the previous sections. A composite system consists of several pattern recognition modules (see Figure 4.8). These individual modules are slightly different from one another and act as individual experts who bring their different skills together to solve a common problem, thus achieving better results than any one expert could. To test the principle of composite systems, three modules are employed. Each individual module is set up, initialised and trained independently. The outputs of the modules are channelled to a decision making module to define the final outputs of the composite system.

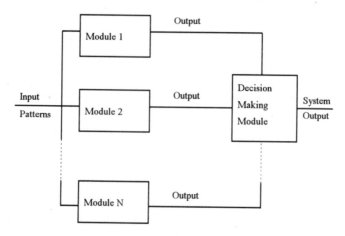

Figure 4.8 General structure of a composite system

In this section, various composite systems utilising Multi-Layer Perceptrons (MLPs), Learning Vector Quantisation (LVQ) networks and conventional (heuristic) pattern recognition systems are described and a comparison of their performances is given. The decision making module is explained first. Details of a composite MLP system, a composite LVQ system and other hybrid systems are then provided.

4.4.1 Decision Making Module

As mentioned above, the decision making module computes the final output of the composite system when an input pattern is presented. Each input pattern to be identified is supplied to all the recognition modules simultaneously. The system outputs are determined on the basis of a "consensus" between the individual modules using the following steps:-

(i) The corresponding outputs of the different modules are summed up, e.g. outputs "a" of modules 1, 2, and 3 are added together.

(ii) If S_{max}, the largest of the sums computed in step (i), exceeds a given threshold τ_1 and all other sums are below τ_1, the system output corresponding to the module outputs that produced S_{max} is set to 1; the other system outputs are set to 0. Otherwise, step (iii) is pursued.

(iii) The corresponding outputs of the modules are added in pairs (e.g. outputs "a" of module 1 and module 2 are added). As the system has three modules, for each output category three sums are then obtained (e.g. the sums Σa_{12}, Σa_{13}, Σa_{23} of outputs "a" of modules 1 and 2, modules 1 and 3 and modules 2 and 3).

(iv) If the overall largest sum Σ_{max} produced in step (iii) is above a threshold τ_2 and all the other sums are below τ_2, then the system output corresponding to the module outputs that produce Σ_{max} is set to 1. The other system outputs are set to 0. Otherwise, step (v) is pursued.

(v) Each system output is set to half of the largest sum produced in step (iv) (that is the average of the largest two corresponding module outputs in the group).

The following examples illustrate how the decision making module operates:

Example 1

	a	b	c	d	e	f
Outputs of module 1	0.01	0.00	0.92	0.00	0.00	0.08
Outputs of module 2	0.00	0.00	0.98	0.00	0.04	0.00
Outputs of module 3	0.00	0.00	0.30	0.00	0.70	0.00
Sums	0.01	0.00	2.20	0.00	0.74	0.08
Threshold τ_1	1.7					
System outputs	0.00	0.00	1.00	0.00	0.00	0.00

Example 2

	a	b	c	d	e	f
Outputs of module 1	0.05	0.00	0.70	0.00	0.70	0.03
Outputs of module 2	0.20	0.00	0.75	0.00	0.20	0.10
Outputs of module 3	0.40	0.00	0.15	0.00	0.15	0.10
Sums	0.65	0.00	1.60	0.00	1.05	0.23
Threshold τ_1	1.7					
Largest sums for group	0.60	0.00	1.45	0.00	0.90	0.20
Threshold τ_2	1.4					
System outputs	0.00	0.00	1.00	0.00	0.00	0.00

Example 3

	a	b	c	d	e	f
Outputs of module 1	0.05	0.00	0.30	0.00	0.70	0.03
Outputs of module 2	0.20	0.00	0.70	0.00	0.20	0.10
Outputs of module 3	0.40	0.00	0.20	0.00	0.15	0.10
Sums	0.65	0.00	1.20	0.00	1.05	0.23
Threshold τ_1	1.7					
Largest sums for group	0.60	0.00	1.00	0.00	0.90	0.20
Threshold τ_2	1.4					
System outputs	0.30	0.00	0.50	0.00	0.45	0.10

4.4.2 Composite MLP System

Structure: The overall structure of the composite MLP pattern recognition system is shown in Figure 4.9. The system consists of 3 independent MLP modules each with the same three-layer architecture as for the MLP system described in Section 4.3. The MLP modules adopted all have 60 neurons in the input layer, 35 neurons in the hidden layer and 6 neurons in the output layer. As before, the input neurons have a unity activation function and the hidden and output neurons have a sigmoidal activation function.

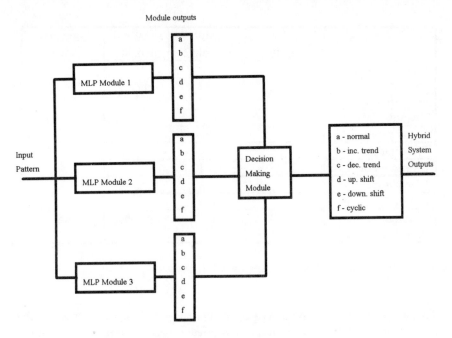

Figure 4.9 Structure of the composite MLP system

Training Data: Three data sets of the same size and constitution as described in section 4.3 are employed to train the network. Two composite pattern recognition systems have been developed, both with identical structures but trained on different data sets. For the first system, all three MLP modules are trained using only one data set. For the second system, each MLP module has to learn a different data set.

Training Method and Results: Both composite systems are trained with a learning rate of 0.3 and a momentum coefficient of 0.8. With the first system, the initial conditions of the networks and order of presentation of the input patterns to the networks are important. For instance, if the initial weights of the connections in the MLP modules are very small, all the modules tend to follow the same solution path after some time, yielding the same result. This is against the design principle of the composite system. Therefore, the initial weights of the connections are set randomly to be reasonably large (between -1 and 1) and the training examples are presented randomly, in order to force each module to adopt a different solution path. A list of the characteristics of this composite system is given in Figure 4.10.

	MLP module 1	MLP module 2	MLP module 3	Composite system
Seed for random number generation	237	712	1991	
Training data		Set 1		
Initial values	-1	to	1	
Presentation		Random		
Learning rate		0.3		
Momentum		0.8		
Learning epochs		200		
Accuracy (threshold = 0.8)	95.2%	94.7%	94.9%	96.8%

Figure 4.10 Features of the first composite MLP system

The system is tested using a data set comprising 1002 unseen patterns. The classification accuracy levels achieved at different stages of the training operation are plotted in Figures 4.11 and 4.12 for different decision module and output selection thresholds. The accuracy levels obtained at the end of 200 training epochs are 96.8%, 95.2%, 94.7% and 94.9% for the composite system and its three components respectively (with an output selection threshold of 0.8 and decision module thresholds $\tau_1=1.7$ and $\tau_2=1.4$).

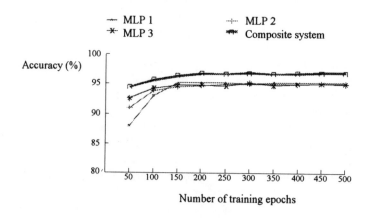

Figure 4.11 Performance of the first composite MLP system ($\tau_1=1.7$; $\tau_2=1.4$; threshold=0.8)

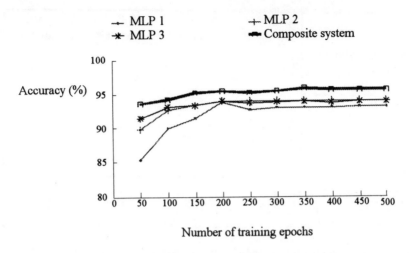

Figure 4.12 Performance of the first composite MLP system (τ_1=2.0; τ_2=1.7; threshold=0.9)

With the second system, the training patterns are presented to the networks sequentially and different training sets are used for each network. Other features of the system are listed in Figure 4.13. The accuracy levels attained by the composite system and its components at different stages of training are measured and plotted for different thresholds (see Figures 4.14 and 4.15). Note that the accuracy level of the composite system reaches 90% after a very short time. For example, at the end of 10 training epochs, the performance of the system is 93.8% while the performances of the individual modules are 73.8%, 86.0% and 80.2% respectively. After 200 training epochs, the system is able to classify 97.1% of the test patterns correctly. The individual accuracy levels of the modules are 94.1%, 94.7% and 94.3% respectively (with an output selection threshold of 0.8 and decision module thresholds τ_1=1.7 and τ_2=1.4.

4.4.3 Composite LVQ System

Structure: The system consists of 3 independent LVQ modules each with the same architecture as for the LVQ system described in section 4.3. Thus, the LVQ modules adopted have 60 neurons in the input layer, 36 neurons in the Kohonen layer and 6 neurons in the output layer.

	MLP module 1	MLP module 2	MLP module 3	Composite system
Seed for random number generation	237	712	1991	
Training data	Set 1	Set 2	Set 3	
Initial values	-1	to	1	
Presentation		Sequential		
Learning rate		0.3		
Momentum		0.8		
Learning epochs		200		
Accuracy (threshold = 0.8)	94.1%	94.7%	94.3%	97.1%

Figure 4.13 Features of the second composite MLP system

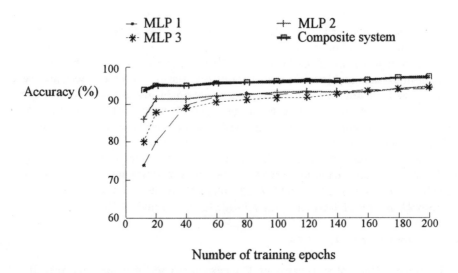

Figure 4.14 Performance of the second composite MLP system
(τ_1=1.7; τ_2=1.4; threshold=0.8)

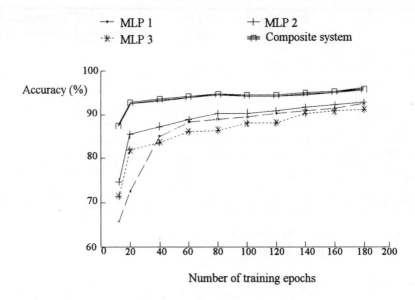

Figure 4.15 Performance of the second composite MLP system (τ_1=2.0; τ_2=1.7; threshold=0.9)

Training data: Each module is given a different training set because using the same data set to train all three modules does not produce a composite system that is distinctly better than its individual components. The data sets employed are the same as those adopted for training the MLP units.

Training Method and Results: The modified LVQ procedure, LVQ-X, is utilised. The initial connection weights are randomly set between -0.1 and 0.1. The training patterns are presented sequentially. The learning rate is first set to 0.05 and decreased monotonically to 0.01 during training. Again, a data set comprising 1002 unseen patterns is used to test the identification accuracy of the composite system.

The three modules require different amounts of training to learn their data sets. The numbers of training epochs for the three modules are 20, 30 and 70 as given in Figure 4.16. The accuracy levels of the composite system and its components for the test data are 99.0%, 97.7%, 97.4% and 88.4%, respectively.

	LVQ module 1	LVQ module 2	LVQ module 3	Composite system
Seed for random number generation	253	124	654	
Initial values	-0.1	to	0.1	
Presentation	Sequential			
Learning rate	0.05 to 0.01			
Coefficient (B)	0.0001			
Coefficient (C)	10			
Learning epochs	20	30	70	
Accuracy (threshold = 0.8)	97.7%	97.4%	88.4%	99.0%

Figure 4.16 Features of the composite LVQ system

4.4.4 Hybrid Composite Systems

Hybrid composite pattern recognition systems are obtained by mixing the MLP, LVQ and heuristic pattern recognition modules.

Composite "MLP + Heuristic" System: The system consists of 2 MLP modules and the heuristic pattern recognition system described in section 4.2. The features of the MLP modules are the same as those employed in the composite MLP system. The networks are trained using either the same data set or separate data sets.

The overall accuracy level of the system is 97.7% when using the same data set to train the networks and 98.2% when separate training data sets are employed. The performances of the individual components of the system are listed in Figure 4.17.

	Same data	Different data
MLP 1	95.2%	95.2%
MLP 2	95.3%	94.3%
Heuristic	94.8%	94.8%
MLP + Heuristic	97.7%	98.2%

Figure 4.17 Performance of the composite "MLP+Heuristic" system

Composite "MLP + LVQ" System: In this composite system, one pattern recognition module based on the LVQ network is combined with two modules based on MLPs. The classification accuracy of the overall system is 97.9% when the three modules are trained with the same data and 98.2% when they learn different data sets. Figure 4.18 shows the individual performances.

	Same data	Different data
MLP 1	95.2%	95.2%
MLP 2	95.3%	94.3%
LVQ	97.7%	97.7%
MLP + LVQ	97.9%	98.2%

Figure 4.18 Performance of the composite "MLP+LVQ" system

Composite "LVQ + MLP" System: This system consists of two LVQ and one MLP pattern recognition modules. The LVQ modules are trained using separate data sets. The MLP module is trained on either one of the data sets used to train a LVQ module or a completely different data set. The overall accuracy level achieved with the best pair of LVQ modules (the best pair are selected from the 3 LVQ modules used in the composite LVQ system of section 4.3) and a MLP module is 99.10% when one of the LVQ networks and the MLP network use the same data set and 98.50% when all networks are trained using different data sets. Individual performances are shown in Figure 4.19.

	Same data	Different data
LVQ 1	97.7%	97.7%
LVQ 2	97.4%	97.4%
MLP	94.7%	94.3%
LVQ + MLP	99.1%	98.5%

Figure 4.19 Performance of the composite "LVQ+MLP" system

Composite "LVQ + Heuristic" System: This system comprises 2 LVQ modules and a heuristic pattern recognition module. Different training data sets are employed to train the LVQ networks. The identification accuracy of the system achieved with the best pair of LVQ modules is 99.5%. Figure 4.20 shows the performance of the system an its components.

LVQ 1	97.7%
LVQ 2	97.2%
Heuristic	94.8%
LVQ + Heuristic	99.5%

Figure 4.20 Performance of the composite "LVQ + Heuristic" system

Composite "MLP + LVQ + Heuristic" System: This final composite system has a MLP module, a LVQ module and a heuristic module. The results obtained are shown in Figure 4.21. It can be seen that the accuracy level achieved by this composite system is 98.9% when the same training set is used for all modules and 99.0% when separate sets are employed.

	Same data	Different data
MLP	95.2%	95.2%
LVQ	97.7%	97.4%
Heuristic	94.8%	94.8%
MLP+LVQ+Heuristic	98.9%	99.0%

Figure 4.21 Performance of the composite "MLP + LVQ + Heuristic" system

4.4.5 Discussion on Using Composite Systems

Different control chart pattern recognition systems have been described in sections 4.2 and 4.3. All these systems are able to classify control chart patterns with a high degree of reliability. Composite pattern recognition systems, however, consistently perform better than their single counterparts. They train faster and identify patterns correctly more often. As mentioned earlier, a possible explanation for this is that, in a composite system, there is synergy between the individual networks which act similarly to a team of specialists solving a common problem. Each specialist is knowledgeable on a different aspect of the problem. The combined effort of the team generally produces better results than would be achievable individually.

The training of neural networks is an optimisation task which is subject to local minima problems. Starting the search for the optimum from different places with different random initial values and using different presentation schemes and different network paradigms help to produce the different specialists needed for a composite system.

A problem with neural networks has been their sometimes unpredictable behaviour and spurious outputs. The adoption of the synergistic arrangement of different networks described here should generally increase reliability because of the cross-checking of the outputs of the individual networks.

A summary of the results obtained with the different pattern recognisers is provided in Figure 4.22. It can be seen that combining the heuristic method with LVQ neural networks yields the most accurate control chart pattern recognition system. Another advantage with using LVQ networks is their training speed which is several orders of magnitude higher than that of MLP networks.

	Same data	Different data
MLP	95.2%	
LVQ	97.7%	
Heuristic	94.8%	
Composite LVQ		99.0%
Composite LVQ + Heuristic		99.5%
Composite MLP	96.8%	97.1%
Composite MLP + LVQ	97.9%	98.2%
Composite LVQ + MLP	99.1%	98.5%
Composite MLP + Heuristic	97.7%	98.2%
Composite MLP + LVQ + Heuristic	98.9%	99.0%

Figure 4.22 Performances of the various pattern recognition systems

4.5 Summary

In this chapter, automatic control chart pattern recognisers utilising heuristic rules as well as neural networks have been presented. Different neural network pattern recognisers based on the MLP and LVQ models have been detailed. Several composite control chart pattern recognition systems involving MLP, LVQ and heuristic pattern recognition techniques have been described. The results obtained with these systems have shown that they are capable of identifying control chart patterns with a very high level of accuracy.

References

Chatfield, C. (1989) *The Analysis of Time Series: An Introduction*, Chapman and Hall, London.

Cheng, C. (1989) *Group Technology and Expert System Concepts Applied to Statistical Process Control in Small Batch Manufacturing*, PhD dissertation, Graduate College, Arizona State University, Tempe, AZ.

Cheng, C. and Hubele, N.F. (1989) A framework for a rule-based deviation recognition system in statistical process control, *Proc. Int. Industrial Engineering Conf. IIE*, Toronto, May 1989, pp. 677-682.

Evans, J.R. and Lindsay, W.M. (1989) *The Management and Control of Quality*, West Publishing Company, St Paul, MN.

Grant, E.L. and Leavenworth, R.S. (1988) *Statistical Quality Control*, McGraw Hill, New York.

Hwarng, H.B. and Hubele, N.F. (1991) X-Bar chart pattern recognition using neural nets, *Proc. 45th Annual Quality Congress. American Society for Quality Control*, Milwaukee, 20-22 May 1991, pp. 884-889.

Kohonen, T. (1990) Self-organizing feature map, *Proc. of the IEEE*, 78(9), 1464-1480.

Kollen, J.F. and Pollack, J.B. (1990) *Backpropagation is Sensitive to Initial Conditions*, Technical Report #90-JK-BPSIC, Laboratory for artificial intelligence research, The Ohio State University.

Lippmann, R.P. (1989) Pattern classification using neural networks, *IEEE Communication Magazine*, pp. 47-64.

Love, P.L. and Simaan, M. (1989) A knowledge-based system for the detection and diagnosis of out-of-control events in manufacturing processes, *Proc. American Control Conf.*, Pittsburgh, PA, vol. 3, June 1989, pp. 2394-2399.

NeuralWare (1993) *NeuralWorks Professional II/Plus and NeuralWorks Explorer*, Reference Guide, NeuralWare Inc., Pittsburgh, PA.

Pham, D.T. and Liu, X. (1995) *Neural Networks for Identification, Prediction and Control*, Springer Verlag, Berlin and London.

Pham, D.T. and Oztemel, E. (1992a) A knowledge-based statistical process control system, *Proc. 2nd Int. Conf. On Automation, Robotics and Computer Vision*, Singapore, 16-18 September 1992, vol. 3, pp. INV-4.2.1-4.2.6.

Pham, D.T. and Oztemel, E. (1992b) Control chart pattern recognition using neural networks, *Int. J. of Systems Engineering*, Special Issue on Neural Networks, 2(4), 256-262.

Pham, D.T. and Oztemel, E. (1994) Control chart pattern recognition using Learning Vector Quantisation networks, *Int. J. of Production Research*, 32(3), 721-729.

Robinson, E.A. (1981) *Least Square Regression Analysis in Terms of Linear Algebra*, Goose Pond Press, Texas.

Rumelhart, D.E., Hinton, G.E. and Williams, R.J. (1986) Learning internal representation by error propagation, *In: Parallel Distributed Processing*, Rumelhart, D.E. and McClelland, J.L. (Eds.), Cambridge MA: MIT press, vol. 1, pp. 312-362.

Western Electric Co. (1985) *Statistical Quality Control Handbook*, 11th Edition, North Carolina, USA.

Simaan, M. and Love, P.L. (1990) Knowledge-based detection of out-of-control outputs in process control, *Proc. 29th IEEE Conf. on Decision and Control*, Honolulu, Hawaii, December 1990, pp. 128-129.

Swift, J.A. (1987) *Development of a Knowledge-based Expert System for Control Chart Pattern Recognition Analysis*, PhD dissertation, Graduate College, Oklahoma State University, Stillwater, Oklahoma.

Wadsworth, H.M., Stephens, K.S. and Godfrey, A.B. (1986*) Modern Methods for Quality Control and Improvement*, Wiley, Singapore.

Chapter 5   Integrated Quality Control Systems

This chapter discusses the integration of expert systems and neural networks to construct intelligent quality control systems. It describes the combination of XPC, the expert Statistical Process Control system, and neural-network-based control chart pattern recognition modules to yield a hybrid system for effective process monitoring, diagnosis and improvement.

5.1 Integration Process

As explained previously, expert systems and neural networks are two knowledge processing tools with contrasting characteristics (see Figure 5.1). They address two different levels of knowledge processing and problem solving. Expert systems are appropriate for expert knowledge (high-level information processing) whereas neural networks are more suitable for perceptual tasks which are difficult to represent by explicit rules (low-level information processing). The integration of an expert system and a neural network would be particularly advantageous for on-line real-time application to statistical process control due to the fact that this problem domain requires both levels of knowledge processing.

	Expert systems	Neural networks
Knowledge	Symbolic	Distributed
Processing task	High level	Low level
Knowledge representation	Explicit	Implicit
Knowledge acquisition	By knowledge engineer (difficult)	Automatic (simpler)
Computation	Complicated (Numbers / symbols / serial)	Simple (Numbers only / parallel)
Time domain	Discrete	Discrete / Continuous

Figure 5.1 General features of expert systems and neural networks

5.1.1 Comparison of Expert Systems and Neural Networks

Expert systems and neural networks have been compared by several researchers [Bounds, 1989; Eliot and Holliday, 1989; Kottai and Bahill, 1989; Sherald, 1989; Miller *et al.*, 1990; Caudill, 1991; Humpert and De Korvin, 1991; Wildberger, 1991; Kasabov, 1996]. An overview of their structural and functional similarities is given in Figures 5.2 and 5.3, respectively.

Expert systems	Neural networks
Premise or conclusion	Node
Rule	Connection
Measures of uncertainty	Weight values
Evidence combining	Combination function
Rule firing	Activation function

Figure 5.2 Structural equivalence between expert systems and neural networks (adapted from [Kuncicky *et al.*, 1991]).

Expert systems	Neural networks
Have a good user interface	Have no user interface (generally)
Have a good explanation capability	Have little or no explanation capacity
Require at least one domain expert for knowledge elicitation	Require many examples
Are difficult to maintain (for systems with large knowledge bases)	Are relatively easy to maintain (maintenance problems do not increase significantly with network size)
Are built through rule development	Are built through training
Perform serial computation (stepwise computation)	Perform parallel computation
Are brittle	Are robust
Are adopted when a domain expert is available for knowledge elicitation and there are not many examples	Are preferred when no domain expert is available for knowledge elicitation and a sufficient number of examples are provided

Figure 5.3 Comparison of expert systems and neural networks [Pham and Oztemel, 1995].

Structurally, the two technologies are akin to each other, even though they use different means to perform information processing tasks. For example, the actions of the *inference engine* in an expert system are similar to the operations of the *nodes* in a neural network. A *connection* in a neural network is equivalent to a *rule* and a *node* represents the *rule premise* or *conclusion*. The values of the connection weights of a neural network correspond to uncertainty measures in the rules of an expert system. The *combination function* of a neuron has a similar effect to *collecting and combining the evidence* to authorise the firing of a rule whereas the task of the *activation function* is analogous to *firing a rule*.

Functionally, expert systems store and process knowledge symbolically. The main assumption on which they are based is that intelligent behaviour may be derived from the knowledge explicitly encoded in them as rules, objects, facts, frames or other types of knowledge structures. This makes expert systems easily understandable but computationally complicated. They perform operations in a sequential manner, allowing easy addition or deletion of rules without affecting the overall structure. Because of their explicit knowledge representation, like a human expert, they can also explain and justify how and why they have arrived at a certain conclusion. However, they cannot generalise beyond the limits of their static knowledge bases.

Neural networks, on the other hand, are modelled on the behaviour and structure of the human brain. They employ parallel distributed knowledge processing, assuming that intelligence can evolve through interactions among the large numbers of processing units which are densely connected to one another. This parallelism enables a neural network to perform very fast information processing. Another feature of neural networks is their ability to generalise and their resilience in the presence of imperfect and noisy data. However, the functioning of a neural network is not open to examination due to its distributed structure which prevents it from operating in a procedural manner. Therefore, it is difficult to explain the reasoning performed by a neural network.

Expert systems have proved highly successful in solving problems that are regularly structured and performing tasks where the rules for decision making are clear and information is reliable. Unfortunately, they are not robust in situations at the boundary of their knowledge domain [Hudson and Cohen, 1991]. In contrast, neural networks can succeed where explicit decision rules are not available and information is incomplete or only partially correct. Incorrect or missing information would create irregular knowledge bases. However, the generalisation capability possessed by neural networks permits them to produce reasonable answers in these situations [Caudill, 1990].

Building an expert system is a labour-intensive and costly task. The knowledge acquisition and refining processes are two major bottlenecks in the development cycle. Usually, at least one expert is required to supply the necessary rules and a knowledge engineer to code them. These rules must be consistent, cogent and correct and must be in a conjunctive form, which does not resemble the human reasoning process. Also, experts do not always think in terms of rules and therefore it could be very difficult for them to articulate their knowledge [Yang and Bhargava, 1990]. In general, the knowledge applied by experts for judgement is implicit in nature as they use their intuition which may be based on many years of experience. The problem in coding this implicit knowledge into explicit rules is that a significant amount of critical information may be lost. Moreover, if there is more than one expert involved and they disagree, then knowledge encoding becomes more difficult for the knowledge engineer. In many cases, it is not possible to produce a sufficiently comprehensive set of rules that can take into account every situation. Even when the desired knowledge has been acquired, it is very difficult to test the consistency of a rule set as there is no simple method to refine and correct a rule base. Additionally, some systems require a large number of rules. This makes them extremely difficult to build and maintain. Finally, the performance of an expert system depends on the depth of the knowledge used, the co-operation of the domain expert and the ability of the knowledge engineer to elicit domain knowledge. All in all, this makes the process of building an expert system very much more time consuming than that of constructing neural networks.

Neural networks can cope with implicit knowledge and are not required to follow the reasoning process adopted by a domain expert in solving a problem. They are trained using past examples which represent the decisions of an expert in dealing with such a problem. As neural networks are able to learn from examples, the knowledge acquisition task is performed automatically. This, however, makes the neural network modification process difficult. Usually, if the input space is changed, new examples are required and often the network must be retrained or reconstructed. This contrasts with the simple facilities of expert systems for adding, deleting and updating rules. Also, structuring a neural network can be a complex task as there is no systematic method for selecting the network topology and associated parameters for a particular problem.

Another difference between expert systems and neural networks concerns learning. It is not easy for an expert system to learn. Although, there are induction procedures developed for this purpose, such as ID3 [Quinlan, 1987] and RULES-3 [Pham and Aksoy, 1995], they are not as efficient as neural network learning algorithms under noisy conditions [Shawlik et al., 1989]. Also, many neural networks can learn gradually as different examples are presented to them. This is a difficult procedure for the majority of expert systems. Finally, a distinguishing aspect of the two tools is that neural networks perform numerical information processing in both the discrete and continuous time domain, whereas expert systems process knowledge only in the discrete time domain, making use of numbers as well as symbols.

5.1.2 Combining Expert Systems and Neural Networks

It is believed that, due to their complementary aspects, combining an expert system with a neural network into a single integrated environment would produce a system that is more capable than its individual components. For example, some features of neural networks, such as their flexibility, parallelism and robustness, would be useful to a rule-based system. An expert system enriched with an adaptive learning ability could greatly simplify the knowledge acquisition task. Parallelism would speed up execution and a generalisation capability could solve the problem of an expert system suddenly failing near the limits of its expertise. On the other hand, certain features of expert systems such as good man-machine interfaces and the ability to provide explanations would make a connectionist system more explicit and user friendly.

An integrated system would be able to handle missing data, uncertain and fuzzy information and situations that only partially match those previously taught to it, producing outputs when an expert system alone fails to do so. A neural network may be used with a rule-based system to obtain weighting factors and membership values or refine certainty factors in a set of rules [Hudson and Cohen, 1991; Kuncicky *et al.*, 1991]. A neural network could also be employed to pre-process data to yield knowledge that could be effectively handled by an expert system.

An integrated system would also be able to generalise, such that non-essential rules could be deleted and new rules added to the rule base. Inconsistent and incorrect rules could be recognised by comparing them with other rules in the knowledge base. The system could also extract rules using neural networks and restore partially lost rules or a damaged rule base [Yang and Bhargava, 1990].

The possible approaches to integration can be grouped into two main categories: functional combination and structural integration.

In a hybrid system of the first category, the neural network and expert system are two separate components as shown in Figure 5.4 [Tirri, 1988; Barker, 1990; Glover *et al.*, 1990; Kasabov, 1990; Handelman *et al.*, 1990; Garcia, 1991; Woodcock *et al.*, 1991]. A large task is decomposed into smaller parts each assigned to that component which is most suited for it. Information freely flows between the expert system and the neural network. Usually, the former is responsible for the overall control of the task due to its ability to provide explanations for the combined reasoning.

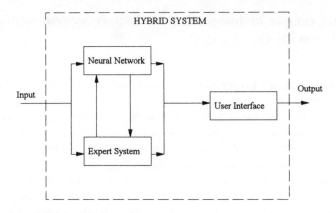

Figure 5.4 Functional combination

In a hybrid system belonging to the second category, usually, the neural network is embedded into the expert system [Gallant, 1988; Samad, 1988; Bradshaw *et al.*, 1989; Fu, 1989; Hall and Romaniuk, 1990; Fu and Fu, 1990; Sun, 1992; Sun, 1994]. Both tools concentrate on solving the same task and all or part of the functionality of the expert system is replaced by the neural network. Often, the neural network is employed as a knowledge base or an inference engine and a knowledge base together (Figure 5.5). In such cases, rules are encoded within the structure of the network. As previously suggested, nodes of the network represent the premise and conclusion of rules with the values of certainty factors equalling the weight values of the connections.

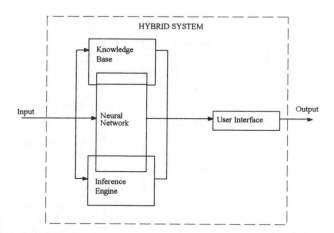

Figure 5.5 Structural integration

5.2 An Example of Integrating an Expert System with Neural Networks for Quality Control

This section describes a hybrid system for quality control developed by Pham and Oztemel [1995]. The system adopts the first method of integration. The expert system used is the XPC program described in Chapter 3. The neural networks are those forming one of the composite pattern recognisers described in Chapter 4. Integration of XPC and a control chart pattern recogniser is desirable because of the need to discover the causes of the abnormal patterns detected by the neural network and to suggest corrective actions. This obviously requires expert diagnostic knowledge. It has been shown that the neural network pattern recognition system and the expert system are individually capable of performing the tasks assigned to them. XPC can track a process to detect out-of-control situations. The composite neural-network-based pattern recognition system is able to predict impending process abnormalities by analysing the patterns exhibited by the control charts.

The components of the hybrid system are illustrated in Figure 5.6. Only the signal pre-processing, on-line monitoring and diagnosis modules of the XPC expert system are shown in this figure. The neural-network-based control chart pattern recognition system comprises three pattern recognition modules: two LVQ neural networks and a heuristic unit. This particular composite system is used as it has the best overall performance. The expert system activates each pattern recognition module from a LEONARDO procedure (Figure 5.7).

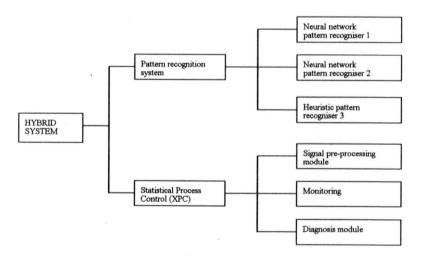

Figure 5.6 Components of the hybrid system

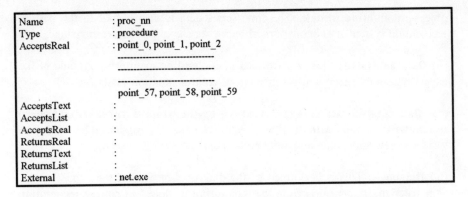

Name	: proc_nn
Type	: procedure
AcceptsReal	: point_0, point_1, point_2

	-------------------------------
	point_57, point_58, point_59
AcceptsText	:
AcceptsList	:
AcceptsReal	:
ReturnsReal	:
ReturnsText	:
ReturnsList	:
External	: net.exe

Figure 5.7 An example of a procedure used to call a neural network module

The values of the process signal (point 0 to point 59) are passed to the neural network program *net.exe* referred to in the procedure (as shown in Figure 5.7) which is then activated. The same procedure is applied to the other pattern recognition modules. The results obtained by them are channelled to the decision making module as explained in Chapter 4. The overall decision of the pattern recognition system about the pattern exhibited is found and sent back to the expert system for interpretation. If just one of the output values from the decision making module is above a threshold, which is taken as 0.7, the type of the pattern is confirmed as the type assigned to that output. Once a pattern is recognised as an abnormal pattern, the knowledge base is searched for possible causes. These are then listed to the operator for correction. If two or more outputs of the pattern recognition system are above the threshold or all outputs are below the threshold, the pattern is not assigned to a unique type.

For example, if the outputs of the pattern recognition system are 0.1, 0.4, 0.0, 0.4, 0.0 and 0.1 then the pattern is regarded as:-

- a normal pattern with 10% certainty;
- an increasing trend with 40% certainty;
- a decreasing trend with 0% certainty;
- an upward shift pattern with 40% certainty;
- a downward shift pattern with 0% certainty;
- a cyclic pattern with 10% certainty.

The operation of the hybrid system is divided into 4 stages:-

(i) **Data collection and pre-processing**: This is implemented by the XPC expert system. The data acquired is supplied on-line to a signal pre-processing procedure in XPC which arranges it into a time series for inputting to the on-line process monitoring module. The time series data is also passed to the pattern recognition system after being normalised in the signal pre-processing stage;

(ii) **Data analysis I:** This is performed by the on-line monitoring module of the expert system to decide if a specified out-of-control situation exists;

(iii) **Data analysis II:** This is carried out by the neural-network-based pattern recogniser in parallel with the first stage of data analysis conducted by the expert system to determine if an abnormal pattern exists;

(iv) **Diagnosis:** This is performed by the diagnosis module of the expert system after receiving information from the monitoring module and pattern recognition system to decide the state of the process and suggest corrective actions if required.

The information flow between the XPC expert system and the neural-network-based pattern recognition system is summarised in Figure 5.8.

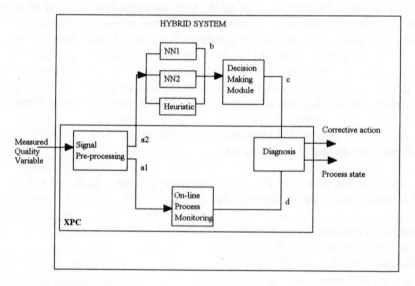

a1 - Raw time series
a2 - Normalised time series
b - Provisional pattern identification
c - Overall pattern identification
d - Process state (provisional information)

Figure 5.8 Information flow in the hybrid system

The management of the hybrid system is the responsibility of the expert system which activates the monitoring and pattern recognition modules each time a new data item is received. When this happens, the process time series is updated by removing the oldest data item from the beginning of the series and adding the new one to the end of the series as previously explained. The time series is then fed to the on-line monitoring and pattern recognition modules simultaneously. The decisions of these two sub-systems are then used by the expert system to perform diagnostic problem solving. Once an abnormal pattern is detected, a warning message is issued, the abnormal data points are isolated and the pattern recognition system is turned off until a new time series signal of the specified length is reformed. Figure 5.9 shows an abnormal pattern detected by the hybrid system when monitoring the temperature in an injection moulding machine. The likely causes of that pattern are given in Figure 5.10. Note that, although the temperature is about to enter an out-of-control state, the expert system alone is unable to predict this and still declares it as being in-control. This is because none of the out-of-control conditions, as covered by the SPC rules embodied in the monitoring module of the expert system, apply and therefore the temperature appears to be normal. However, the hybrid system successfully predicts an impending abnormal situation using its neural-network-based pattern recognition module to analyse the patterns in the temperature over a longer time scale than covered by the SPC rules.

5.3 Summary

This chapter has discussed the integration of expert systems and neural networks to provide improved quality assurance tools. The chapter has described a hybrid system that integrates an expert system for Statistical Process Control with a neural-network-based control chart pattern recognition unit. The hybrid system benefits both from the good user interface of its expert system component and from the robust predictive ability of its neural-network-based component.

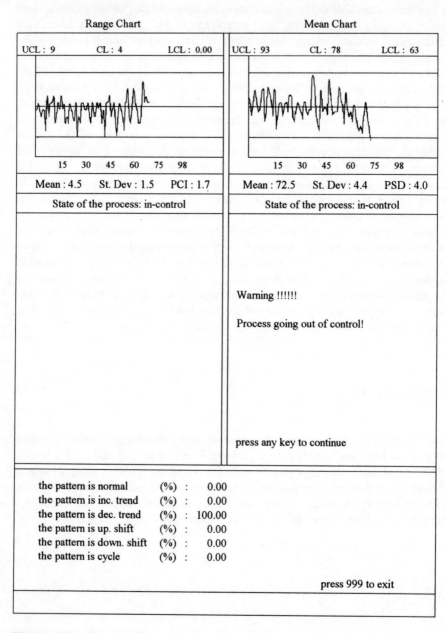

Figure 5.9 An example of a decreasing trend not detected by XPC but recognised by the hybrid system

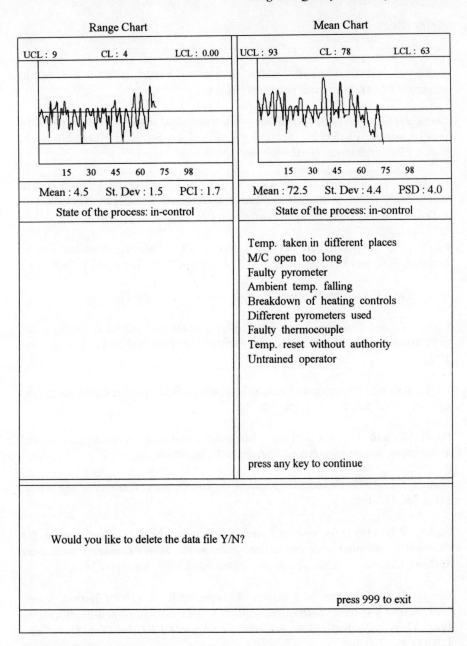

Figure 5.10 Causes of the pattern shown in Figure 5.9.

References

Barker, D. (1990) Financial health: Integrating expert systems and neural networks, *PC AI*, May/June, pp. 24-27: 62-64.

Bounds, D.G. (1989) Expert systems and connectionist networks, *Connectionism in Perspective*, Pfeifer, R., Schreter, Z., Fogelman-Soulie, F., and Steels, L. (eds:), Elsevier Science, North-Holland, pp. 277-282.

Bradshaw, G., Fozzard, R. and Ceci, L. (1989) A connectionist expert system that actually works, *Advances in Neural Information Processing Systems*, vol. 1, D.S. Touretzky (ed.), Morgan Kaufmann, Los Altos, CA, pp. 248-255.

Caudill, M. (1990) Expert networks, *Neural PC Tools: A Practical Guide*, Eberhart, R.C. and Dobbins, R.W. (eds.), Academic Press, London, pp. 188-214.

Caudill, M. (1991) Expert networks, *Byte*, October, pp. 108-116.

Eliot, L.B. and Holliday, F. (1989) Expert systems and neural networks - An experimental study of methodological expertise, *Neural Networks*, 1(2), pp. 96-106.

Fu, L.M. (1989) Integration of neural heuristics into knowledge-based inference, *Connectionist Science*, 1(3), pp. 325-340.

Fu, L.M. and Fu, L.C. (1990) Mapping rule-based systems into neural architecture, *Knowledge-Based Systems*, 3(3), pp. 48-58.

Gallant, S.I. (1988) Connectionist expert system, *Communications of the ACM*, 31(2), pp. 152-169.

Garcia, R.P. (1991) A rule and neural net based hybrid expert system for electrical transformer price estimation, *Proc. of the World Congress on Expert Systems*, Orlando, Florida, vol. 1, 16-19 December 1991, pp. 514-521.

Glover, C.W., Silliman, M., Walker, M. and Spelt, P. (1990) Hybrid neural network and rule-based pattern recognition system capable of self-modification, *Proc. SPIE - Applications of Artificial Intelligence VIII, Int. Soc. Opt. Eng.*, 1293(1), pp. 290-300.

Hall, L.O. and Romaniuk, S.G. (1990) A hybrid connectionist learning system, *Proc. of AAAI-90, 8th National Conf., American Society of Artificial Intelligence*, Boston, MA, pp. 783-788.

Handelman, D.A., Lane, S.H. and Gelfand, J.J. (1990) Integrating neural networks and knowledge-based systems for intelligent robotic control, *IEEE Control Systems Magazine*, April, pp. 77-87.

Hudson, D.L. and Cohen, M.E. (1991) Combinations of rule-based and connectionist expert systems, *Microcomputer Applications,* 10(2), pp. 36-41.

Humpert, B. and De Korvin, A. (1991) Optimisation of expert systems with neural networks, *Int. J. of Mod. Physics: C*, 2(1), pp. 86-104.

Kasabov, N.K. (1990) Hybrid connectionist rule-based systems, *Artificial Intelligence IV, Methodology, Systems and Explanations*, Jorrand, P.H. and Squrev, V. (eds.), Elsevier, North-Holland, pp. 227-235.

Kasabov, N.K. (1996) *Neural Networks, Fuzzy Systems and Knowledge Engineering*, MIT Press, Cambridge, MA.

Kottai, R.M. and Bahill, A.T. (1989) Expert systems made with neural networks, *Int. J. of Neural Networks,* 1(4), pp. 211-226.

Kuncicky, D., Hruska, S.I. and Lacher, R.C. (1991) *Hybrid Systems: The Equivalence of Rule-Based Expert System and Artificial Neural Network Inference*, Dept. of Computer Science, Florida State University.

Miller, R.K., Walker, T.C. and Ryan, R.M. (1990) *Neural Net Applications and Products*, SEAI Publications and Graeme Publications, USA, Chapter 8, pp. 144-150.

Pham, D.T. and Aksoy, M.S. (1995) A New Algorithm for Inductive Learning, *J. of Systems Engineering*, 5(2), 115-122.

Pham, D.T. and Oztemel, E. (1995) An integrated neural network and expert system tool for statistical process control, *Proc. I Mech E, Part B: J. of Engineering Manufacture*, vol. 209, 91-97.

Quinlan, J.R. (1987) Simplifying decision trees, *Int. J. of Man Machine Studies*, vol. 27, pp. 221-234.

Samad, T. (1988) Towards connectionist rule-based systems, *IEEE Int. Joint Conf. on Neural Networks*, San Diego, California, 24-27 July, pp.II-525 - II-532.

Shawlik, J.W., Mooney, R.J. and Towell, G.G. (1989) *Symbolic and Neural Learning Algorithms: An Experimental Comparison*, Technical Report #857, Computer Science Dept., University of Wisconsin, Madison.

Sherald, M. (1989) Neural networks versus expert systems: Is there a room for both?, *PC AI*, July/August, pp. 10-15.

Sun, R. (1992) Connectionist models of rule-based reasoning, *AISB Quarterly*, vol. 79, pp. 21-24.

Sun, R, (1994) *Integrating Rules and Connectionism for Robust Commonsense Reasoning*, Wiley, New York, NY.

Tirri, H. (1988) Applying neural computing to expert system design: Coping with complex sensory data and attribute selection, *3rd Int. Conf. on Foundations of Data Organisation and Algorithms,* Paris, France, June, pp. 474-488.

Wildberger, A.M. (1991) Comparison of neural network and expert system technologies, *Artificial Intelligence and Simulation*, Uttamsingh, R.J. and Wildberger, A.M. (eds.), Los Altos, California, The Soc. for Comp. Simulation, USA, simulation series, 23(4), pp. 152-158.

Woodcock, N., Hallam, N.J., Picton, P.D. and Hopgood, A.A (1991) Interpretation of ultrasonic images of weld defects using a hybrid system, *Proc. Neuro-Nimes'91, 4th Int. Conf. on Neural Networks and their Applications*, Nimes, 4-8 Nov, pp. 593-605.

Yang, Q. and Bhargava, V.K. (1990) Building expert systems by a modified perceptron network with rule transfer algorithm, *Int. Joint Conf. on Neural Networks,* San Diego, 2, 17-21 June, pp. II-77 - II-82.

Chapter 6   Experimental Quality Design

The quality of a product normally depends on the parameters (variables) which govern the behaviour of the process for manufacturing it. Some of these parameters, such as the temperature in a furnace or the speed and feed rate in a metal removal process, may be controlled by the manufacturer/designer. They are usually referred to as the *control parameters* in a manufacturing process. However, many of them, such as environmental factors, may not be under the control of a designer. They are often called *noise variables*. This chapter discusses simple but effective methods for dealing with these variables to improve quality.

6.1 Taguchi Experimental Design

A noise variable could be *external* or *internal*. External sources of noise are variables external to a product, such as variation in environmental conditions and human factors in operating a manufacturing process. An internal source of noise is the deviation of the actual characteristics of a product from the nominal settings due to manufacturing imperfections or product deterioration. Taguchi [1986].has suggested 3 stages of design to minimise the effects of noise variables (see Figure 6.1). These are:-

System design. This design stage is also called 'primary' design. It applies basic scientific and engineering knowledge to produce a prototype design bearing in mind customer needs and manufacturing requirements. Past experience, intuition, and existing knowledge are the basic sources of information on which system design relies. At the end of system design, materials, parts, product parameters, production equipment and process factor values are identified.

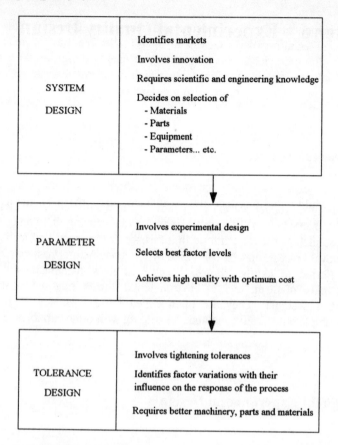

Figure 6.1 Three stages of quality design

Parameter design. This design stage is also called 'robust' design. Its main aim is to reduce costs and improve quality. This is achieved through deriving optimum parameter settings using statistical techniques and experiments. Parameter design is the most important stage in the design of quality products. The remainder of this section is therefore devoted to a detailed explanation of this design stage.

Allowance or tolerance design. If the noise variables cannot be reduced by good parameter design, Taguchi suggests tolerance design where the tolerances of the product components are reconsidered and redesigned (possibly reducing the tolerances of certain crucial factors in a cost effective way or relaxing the tolerances of non-crucial factors). This stage of quality design might necessitate new materials, equipment or components, which in turn increases costs. It should therefore be noted that tolerance design is applied only if parameter design has failed.

6.1.1 Parameter Design

Parameter design is an important method for improving product manufacturability and product life, which in turn increases manufacturing process stability. It is defined as the operation of choosing settings for product and process parameters to reduce variations in the manufacture of the product [Taguchi, 1986]. The main aim of this type of design is to arrange the settings of the control parameters in such a way that the effect of noise variables is minimised. A schematic representation of parameter design is shown in Figure 6.2.

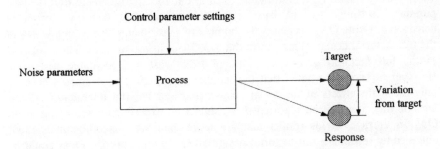

Figure 6.2 Schematic representation of parameter design

The main steps in parameter or robust design are:-

(i) formulating the problem and separating the factors: This step includes the definition of the objectives and quality characteristics which affect those objectives. Control parameters and noise variables should also be listed. This step usually involves brainstorming by a group of engineers. Once the quality characteristics are identified, it is important to define the variables (factors) and their interactions that influence quality. The term *interaction* between two factors is used to describe a condition in which the effect of one factor upon the observed response value is dependent on the condition of the other. Each factor is identified with its possible parameter settings. This means that the different levels of each factor or variable with their corresponding values need to be listed in order to discover the best factor-level combinations.

(ii) planning the experiments or experimental set up: In this step, the plan for changing the order of the parameter settings in the experiments designed to determine the optimum setting is prepared. To do this, Taguchi suggests the use of *orthogonal arrays* (OAs). These arrays are tables of integers where columns are pairwise balanced. This means that every level of a factor appears the same number of times in any two columns of a table. Examples of several different orthogonal arrays are given in Figure 6.3. These arrays allow the factors to have different test levels and ensure a minimum number of experiments [Shoemaker and Kacker, 1988]. The design of experiments using these arrays is often called fractional factorial design because they are a subset of experiments from a full factorial design. For example, if there are 5 factors with 2 levels each, 32 (2^5) possible experiments are required for a full factorial design. However, using the OA_{16} orthogonal array, it is sufficient to carry out 16 experiments to find the best parameter settings. In this case, only half of the full factorial design is considered. Using OA_8 decreases the number of experiments to 8 which is 1/4 of the full factorial design. This shows the benefits of using an OA when compared to the full factorial design in terms of time, cost and available resources. However, it should be noted that the number of experiments decreases because only the main effects of factors are considered and not the interaction effects. Despite this, it has been reported that conducting experiments according to an OA, on average, yields results that are more than 90% as efficient as those obtained by running a full factorial design [ASI, 1990]. This 10% loss could be justified when compared to the costs of the experiments.

Test Number	Column 1	Column 2	Column 3
1	1	1	1
2	1	2	2
3	2	1	2
4	2	2	1

Figure 6.3 a) OA_4 array

Test Number	Column						
	1	2	3	4	5	6	7
1	1	1	1	1	1	1	1
2	1	1	1	2	2	2	2
3	1	2	2	1	1	2	2
4	1	2	2	2	2	1	1
5	2	1	2	1	2	1	2
6	2	1	2	2	1	2	1
7	2	2	1	1	2	2	1
8	2	2	1	2	1	1	2

Figure 6.3 b) OA8 array

Test Number	Column			
	1	2	3	4
1	1	1	1	1
2	1	2	2	2
3	1	3	3	3
4	2	1	2	3
5	2	2	3	1
6	2	3	1	2
7	3	1	3	2
8	3	2	1	3
9	3	3	2	1

Figure 6.3 c) OA9 array

Test Number	Column														
	1	2	3	4	5	6	7	8	9	10	11	12	13	14	15
1	1	1	1	1	1	1	1	1	1	1	1	1	1	1	1
2	1	1	1	1	1	1	1	2	2	2	2	2	2	2	2
3	1	1	1	2	2	2	2	1	1	1	1	2	2	2	2
4	1	1	1	2	2	2	2	2	2	2	2	1	1	1	1
5	1	2	2	1	1	2	2	1	1	2	2	1	1	2	2
6	1	2	2	1	1	2	2	2	2	1	1	2	2	1	1
7	1	2	2	2	2	1	1	1	1	2	2	2	2	1	1
8	1	2	2	2	2	1	1	2	2	1	1	1	1	2	2
9	2	1	2	1	2	1	2	1	2	1	2	1	2	1	2
10	2	1	2	1	2	1	2	2	1	2	1	2	1	2	1
11	2	1	2	2	1	2	1	1	2	1	2	2	1	2	1
12	2	1	2	2	1	2	1	2	1	2	1	1	2	1	2
13	2	2	1	1	2	2	1	1	2	2	1	1	2	2	1
14	2	2	1	1	2	2	1	2	1	1	2	2	1	1	2
15	2	2	1	2	1	1	2	1	2	2	1	2	1	1	2
16	2	2	1	2	1	1	2	2	1	1	2	1	2	2	1

Figure 6.3 d) OA16 array

Test Number	Column							
	1	2	3	4	5	6	7	8
1	1	1	1	1	1	1	1	1
2	1	1	2	2	2	2	2	2
3	1	1	3	3	3	3	3	3
4	1	2	1	1	2	2	3	3
5	1	2	2	2	3	3	1	1
6	1	2	3	3	1	1	2	2
7	1	3	1	2	1	3	2	3
8	1	3	2	3	2	1	3	1
9	1	3	3	1	3	2	1	2
10	2	1	1	3	3	2	2	1
11	2	1	2	1	1	3	3	2
12	2	1	3	2	2	1	1	3
13	2	2	1	2	3	1	3	2
14	2	2	2	3	1	2	1	3
15	2	2	3	1	2	3	2	1
16	2	3	1	3	2	3	1	2
17	2	3	2	1	3	1	2	3
18	2	3	3	2	1	2	3	1

Figure 6.3 e) OA18 array

To implement an orthogonal array, control parameters (factors) are assigned to the columns of the table and the actual levels of these factors to the corresponding integers in these columns.

The noise plan is the second phase of the experimental set up. It is important to introduce noise variables directly in the experiments. For example, although humidity is a noise variable in some processes and cannot be controlled directly, the experimenter can carry out experiments under different humidity levels. This means using the humidity variable as a direct component of the experiment. However, this is not always possible. The user should find a way to measure changes in these variables. One way could be just to let the changes happen and simply take the measurements at different levels.

(iii) analysing the results (measuring the performance): This step mainly involves measuring the performance obtained with the improved settings of the controllable variables. Taguchi suggests a *signal-to-noise-ratio* (SNR) for measuring the noise performance which, in turn, yields the performance achieved with the improved settings.

This ratio is given by:-

$$SNR = -10 \log_{10}(MSD) \tag{6.1}$$

where MSD is the mean squared deviation from the target value of the quality characteristic. Logothetis and Wynn [1989] have presented and analysed variants of this equation. In essence, the SNR shows which factor has more effect on the response value; the higher the SNR, the more effect the factor has.

Measuring the prediction capability is another way of assessing the performance of the system. Prediction is made with parameter settings which are not used in the actual experimental stage (i.e. not listed in the OA). The predicted response value Y' for parameter setting "A_1 B_1 C_2 ..." is given by:-

$$Y' = \bar{y} + (\overline{A_1} - \bar{y}) + (\overline{B_1} - \bar{y}) + (\overline{C_2} - \bar{y}) + \ldots\ldots \tag{6.2}$$

where $\overline{A_1}$ is the average response value corresponding to the first level of factor A in the orthogonal array and $\overline{B_1}$ and $\overline{C_2}$ are defined similarly. $\bar{y}$ is the average of all response values of the orthogonal array. Once the predicted value is calculated, it can be compared with the actual value obtained by experimentation and the error can be measured to find the performance of the system.

(iv) confirming the experiment: This step is taken when a suitable parameter setting has been suggested. The aim is to confirm the results by conducting a limited experiment. If improved results are indeed obtained then the new design is adopted. If the objectives are not met, the first step needs to be retaken (possibly after a tolerance redesign).

6.1.2 Knowledge-Based Systems for Taguchi Experimental Design

With the availability of knowledge-based tools, it is possible to devise computer systems which can assist users to design complex experiments quickly and analyse the resulting data correctly.

Designing experiments to establish the best combination of parameter settings for a process can be a difficult task as the choice of factors and levels is influenced by technical considerations, time constraints, available resources and the budget required for conducting experiments. Two of the main problems in a Taguchi experimental design process are (i) the selection of a suitable OA and (ii) the assignment of factors and interactions to the appropriate columns of the selected OA.

The first problem, the selection of OAs, requires more attention. There is an increasing need for intelligent selectors which could accept factors, factor levels and degrees of freedom [Roy, 1990] and suggest the best type of OA for a given application in order to minimise the cost and the number of experiments required.

Figure 6.4 shows the main features of standard orthogonal arrays. As mentioned previously, each OA can accommodate a certain number of factors. The allowed number of levels for each factor is different with different arrays. OA_4, for example, can only accommodate 3 factors and all must have 2 levels. However, with minor modifications, it is possible to use the same OA in different ways. For example, OA_8 can cope with 7 factors having 2 levels each but it can also be used to study one 4-level factor and four 2-level factors. However, it should be noted that it is sometimes not possible to have a match between the levels of the factors and the columns of the OAs. For example, if the problem requires a factor with 2 levels and another with 4 levels, none of the available OAs can cope with this situation. The designer needs to solve this type of problem before selecting an OA. One solution is to combine the columns of the array. This allows a 4-level factor to be represented by two factors each having 2 levels. The procedure is simply to consider a 4-level factor as if it were two 2-level factors that interact. This can be illustrated using OA_8. The 4-level factor A can be assigned to columns 1 and 2. The levels of A are then obtained from the ordered pairs of these two columns. That is, (1,1) becomes the first level of A ("A_1"), (1,2) the second level of A ("A_2"), (2,1) the third level of A ("A_3") and (2,2) the fourth level of A ("A_4"). The third column must be set aside because it indicates the interactions between columns 1 and 2. This makes the third column confounded with the main effect of A. An alternative method is to introduce so-called "dummy levels". This is done by repeating one of the existing levels as many times as required. If the problem has a 2-level factor but the OA requires a 3-level factor then either one of the levels of that factor can be duplicated and used as the third level of that factor. The above examples show the complexity of OA selection and imply the possible need for intelligent knowledge-based systems due to the amount of process and design experience involved. The requirement for knowledge-based systems can further be justified by examining other aspects of OAs.

Orthogonal array	Number of experiments	Number of factors	Maximum number of factors at level							
			2	3	4	5	6	7	8	9
OA$_4$	4	3	3	-	-	-	-	-	-	-
OA$_8$	8	7	7	-	-	-	-	-	-	-
--		5	4	-	1	-	-	-	-	-
OA$_9$	9	4	-	4	-	-	-	-	-	-
OA$_{12}$	12	11	11	-	-	-	-	-	-	-
OA$_{16}$	16	15	15	-	-	-	-	-	-	-
--		5	-	-	5	-	-	-	-	-
--		7	3	-	4	-	-	-	-	-
--		9	8	-	-	-	-	1	-	
OA$_{18}$	18	8	1	7	-	-	-	-	-	-
OA$_{25}$	25	6	-	-	-	6	-	-	-	-
OA$_{27}$	27	13	-	13	-	-	-	-	-	-
--		10	-	9	-	-	-	-	-	1
OA$_{32}$	32	31	31	-	-	-	-	-	-	-
--		10	1	-	9	-	-	-	-	-
OA$_{36}$	36	23	11	12	-	-	-	-	-	-
--		16	3	13	-	-	-	-	-	-
OA$_{50}$	50	12	1	-	-	11	-	-	-	-
OA$_{54}$	54	26	1	25	-	-	-	-	-	-
OA$_{64}$	64	63	63	-	-	-	-	-	-	-
		21	-	-	21	-	-	-	-	-
OA$_{81}$	81	40	-	40	-	-	-	-	-	-

Figure 6.4 Features of orthogonal arrays

The total number of degrees of freedom also affects the selection process. The number is determined by summing up the individual degrees of freedom for the factors and interactions. Note that the number of degrees of freedom for a factor is one less than its number of levels and the number of degrees of freedom for an interaction is the product of their individual number of degrees of freedom. An OA can only handle up to a given number of degrees of freedom (usually one less than the number of rows in the array). This shows the effect of degrees of freedom in the selection process.

The following is a general rule for OA selection:-

> IF process requires 2 or 3 factor levels
> AND array is able to handle the total number of degrees of freedom
> for the given problem
> > THEN select an orthogonal array which:-
> > > has 2 and 3 level columns;
> > OR   has 3 level columns only;
> > OR   has 2 level columns only.

Special rules may be required according to the characteristics of the problem. When special cases arise and none of the OAs can be employed, the modification techniques mentioned above can be adopted. Taguchi suggests the use of linear graphs to match a problem to one of the standard OAs. OAs have standard linear graphs and the problem needs to be matched to one of those graphs. Lee *et al.* [1989] have suggested an inference mechanism for pattern matching and Tsui [1988] has reviewed several other approaches for this purpose.

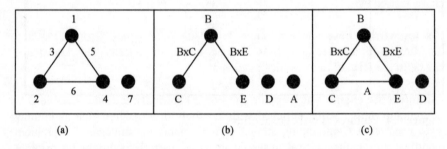

(a)   (b)   (c)

Figure 6.5   (a) standard linear graph for OA$_8$ (b) linear graph of a particular problem (c) graph for the problem fitted to that of OA$_8$

Once the problem has been modified and fitted to a standard OA, then it is easy for the designer to assign the factors to the various columns. The linear graph of that array indicates which factor should be assigned to which column. For example, if there are 5 factors ("A" to "E") with 2 levels each and there are two interactions (BxC) and (BxE) then the problem could be fitted to OA$_8$. The standard linear graph for OA$_8$, the graph for the given problem and the graph for the problem after it has been fitted to that of OA$_8$ are depicted in Figures 6.5(a),

(b) and (c) respectively. Links 3, 5 and 6 in Figure 6.5(a) (or columns 3, 5 and 6 in OA_8) are normally reserved for interactions. Comparing Figures 6.5(a) and (b) shows that factors B, C and E should be assigned to columns 1, 2 and 4 and interactions BxC and BxE to columns 3 and 5. There are no further interactions to cater for. Hence, either factor A or D can be assigned to column 6 and the remaining factor to column 7 as shown in Figure 6.5(c).

Another example of how to assign factors to the columns of an OA is summarised in the following rule:-

> IF there are 4 factors (F_1 to F_4)
> AND there are 3 two-way interactions between F_1, F_2 and F_3
> AND OA_8 is found to be suitable
>> THEN assign F_1, F_2, F_3 and F_4 to the 1st, 2nd, 4th and 7th columns and F_1xF_2, F_1xF_3 and F_2xF_3 to the 3rd, 5th and 6th columns respectively.

A knowledge-based system for assisting the task of experimental design is characterised by:-

a knowledge base which contains knowledge about the existing OAs and modification procedures as well as a variety of design patterns. This knowledge can be in the form of rules as illustrated above.

an inference engine which applies the stored knowledge to find a suitable OA. Inferencing must also be enriched with a variety of strategies such as dummy level generation, confounding strategies and combining strategies for problem modification. That is, the system should advise the user on modifying the original problem whenever appropriate. For example, if a factor or an interaction causes the selection of a large OA, then the system should suggest that the user ignores non-crucial factors or interactions. The inference mechanism should also be able to perform pattern matching in order to fit the problem graph to one of those stored in the knowledge base to facilitate factor assignment.

a user interface module which incorporates explanation facilities and provides suggestions, recommendations and trade-offs in a design process.

6.2   Neural Networks for the Design of Experiments

As explained in the previous section, to identify the optimal design, Taguchi avoids the need for running a full factorial design by using OAs. However, the application of fractional factorial design requires assumptions to be made in identifying the optimal design parameters and consequently leads to uncertainties in the results. For example, the interaction effects which are usually not known to the experimenter are assumed to be negligible if all columns of the array are used by the individual factors. If interactions exist, Taguchi suggests that the factors and factor levels should be carefully chosen to minimise their effects. The designer cannot always achieve this minimisation and this can yield misleading results [Pignatiello and Ramberg, 1991-92]. Moreover, many of the relationships between the inputs and outputs (responses) of the process are non-linear. The more complex the relationship, the harder it is to find the optimum design using the Taguchi method. It is therefore necessary to consider ways to overcome these shortcomings.

Artificial neural networks are an attractive alternative for the task of experimental design. Besides having prediction capabilities, they can also easily separate complex non-linear patterns which cause difficulties for the Taguchi method. An example of the use of a Multi-Layer Perceptron (MLP) to select the best combination of design parameters for a robotic sensor has been described by Rowlands *et al.* [1996]. The principal operation of the sensor is detailed in [Pham and Dissanayake, 1991]. The device consists of a circular platform mounted on the shaft of a motor (Figure 6.6). The platform/motor assembly is constrained to vibrate about an axis perpendicular to the motor shaft, the axis of vibration and the motor shaft being separated by a distance e. The platform supports the part whose position and orientation are to be found by the sensor. The part is firmly held on the platform by a magnet or by vacuum suction. Its position and orientation on the platform are computed from the frequencies of vibration of the platform/motor assembly, with the platform (and thus the part) brought to different locations by the motor. Important factors and their levels are listed in Figure 6.7.

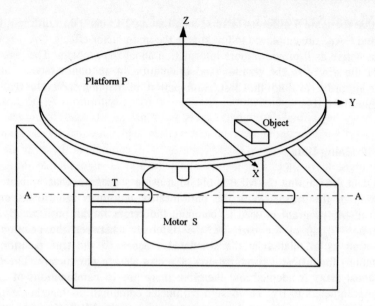

Figure 6.6 Mechanical arrangement of the sensor

Factors		Low (1)	High (2)
A. Stiffness of the spring restraining the movement of the vibrating assembly	(Nm/rad)	10	100
B. Eccentricity of the platform (distance e)	(mm)	2	5
C. Position of the object	(mm)	5	10
D. Mass of the object	(kg)	0.25	0.50
E. Accuracy of the vibration frequency measurement	(%)	1	10

Figure 6.7 Factors and their levels which affect the computation of the position and orientation of an object

Note that there are 5 different factors, each having 2 levels. Two different OAs, OA_8 and OA_{16}, are employed to investigate the main factor effects. OA_{16} is more representative as it provides more information about the problem. The task is to predict the errors in the position and orientation of an object located on the sensor in order to find the best combination of parameter levels that will minimise the errors.

6.2.1 Training Data

The data for training the MLP is obtained using a program that simulates the operation of the sensor. With the part located at known coordinates on the platform, the program is used to compute the errors in the position (d_p) and orientation (d_{gam}) of the part (i.e. the differences between the position and orientation as calculated by the simulated sensor and the true position and orientation) for different combinations of sensor design parameters. The OA_{16} orthogonal array is adopted and therefore there are 16 combinations of sensor design parameters to try. The design parameter combinations together with the corresponding errors d_p and d_{gam} are used to train the MLP. The use of data represented by an orthogonal matrix as the training set for a MLP has also been described in [Roy *et al.*, 1993].

As there are in total 32 possible combinations (due to the 5 two-level design parameters), there are 16 combinations not in the training data set. These are employed to test the prediction performance of the trained MLP.

6.2.2 Structure of the Network

The MLP network has 3 layers: an input layer with 5 neurons corresponding to the selected factors, a hidden layer with 20 neurons and an output layer with 2 neurons giving the position (d_p) and orientation (d_{gam}) errors (see Figure 6.8).

6.2.3 Network Training and Results

The learning rate is initially set at 0.6 and then reduced to 0.05 during training. The momentum coefficient is fixed at 0.8.

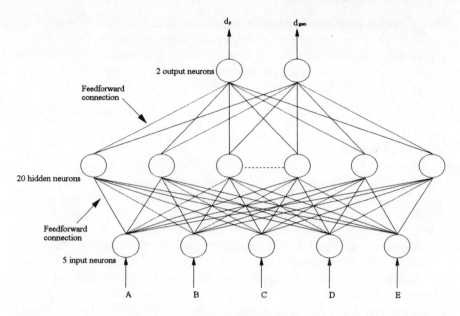

Figure 6.8 Topology of the network used

The neural network output error percentage is calculated as:-

$$\text{Error (\%)} = \frac{\text{Actual value} - \text{Predicted value}}{\text{Actual value}} \times 100 \qquad (6.3)$$

Training is stopped when it is no longer possible to reduce the network error. Figure 6.9 shows the decrease in the total error level during learning for the output neuron representing the error in position d_p. The d_p predictions made by both the Taguchi method and the neural network method are plotted in Figure 6.10. This figure shows that the neural network performs better prediction than the Taguchi method. The error analysis and prediction results for d_{gam} are similar to those for d_p. Average errors are calculated by summing all the error percentages and dividing the results by the total number of errors summed. The average d_p and d_{gam} prediction errors are approximately 9% and 12% for the MLP and 139% and 69% for the Taguchi method respectively.

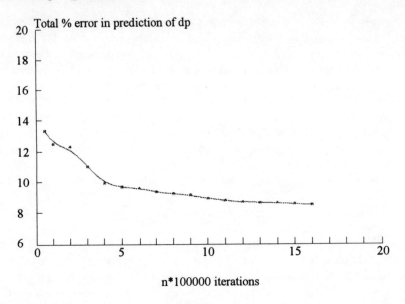

Figure 6.9 Decrease in the error during learning

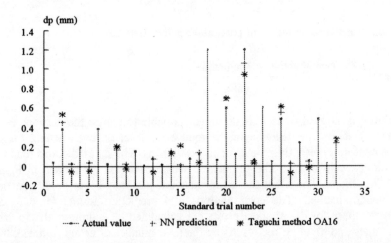

Figure 6.10 Predicted values for d_p obtained with the neural network and the Taguchi method

From Figure 6.10, in nearly all cases, the response of the neural network is much closer to the actual response of the simulated sensor than the Taguchi approach. The factor interactions have less influence on the neural network than on the Taguchi results. This conclusion has also been confirmed by Schmerr *et al.* [1991]. However, the main aim here is not to demonstrate the superiority of neural networks over the Taguchi method but to show the capabilities of neural networks in handling experimental design. Further analysis performed on neural networks has revealed that they could be applied to training sets which do not even satisfy the properties of any orthogonal array. However, neural networks need a good training set, which would be difficult to obtain in some cases. It should be noted that although neural networks do not directly give the best parameter setting, they can be employed as part of a design procedure to locate that setting.

It is clearly stated in the literature that a full factorial design is necessary to find the optimum parameter values of a system. This is an extremely costly and time consuming task. Training neural networks with an OA creates a system which can simulate the full factorial design by predicting response values. This prediction capability is as high as 90-95% which is remarkable considering the cost savings which can be achieved. Once a better solution is suggested by the neural network, it should be confirmed by performing an actual experiment. This opens a new area of research for designers to integrate neural networks with the Taguchi method (or possibly a knowledge-based Taguchi system). It is believed that this integration would better serve the needs of the experimenter for producing higher quality designs.

6.3 Summary

This chapter has discussed the planning of experiments to establish the optimum setting for the parameters of a process or product to ensure that high quality is "designed in". The Taguchi method of experimental quality design has been described. The use of knowledge-based techniques to assist the selection of Taguchi orthogonal arrays has been outlined. An example of the application of a neural network to experimental design has been presented.

References

ASI (1990) *Taguchi Methods*, American Suppliers Institute Inc.

Lee, N.S., Phadke, M.S. and Keny, R. (1989) An expert system for experimental design in off-line quality control, *Expert Systems*, 6(4), 234-249.

Logothetis, N. and Wynn, H.P. (1989) *Quality Through Design: Experimental Design, Off-line Quality Control and Taguchi's Contributions*, Clarendon Press, Oxford.

Pham, D.T. and Dissanayake, M.W.G. (1991) A vibratory sensor for locating parts, *Robotica*, 10(4), 289-302.

Pignatiello, J.J. and Ramberg, J.S. (1991-92) Top ten triumphs and tragedies of Genichi Taguchi, *Quality Engineering*, 4(2), 211-225.

Rowlands, H., Packianather, M.S. and Oztemel, E. (1996) Using artificial neural networks for experimental design in off-line quality control, *Journal of Systems Engineering*, 6(1) (in press).

Roy, R. (1990) *A Primer on the Taguchi Method,* Competitive Manufacturing Series, Van Nostrand Reinhold, New York.

Roy, R., Cave, P. and Parmee, I. (1993) *Orthogonal matrix as a training set for backpropagation neural networks*, Research Report, Plymouth Engineering Design Centre, Plymouth, UK.

Schmerr, L.W., Nugen, S.M. and Forouraghi, B. (1991) Planning robust design experiments using neural networks and Taguchi methods, *Proc. of the Artificial Neural Networks in Engineering (ANNIE'91) Conference*, 10-13 Nov 1991, St Louis, Missouri, USA, pp. 829-834.

Shoemaker, A.C. and Kacker, R.N. (1988) A methodology for planning experiments in robust product and process design, *Quality and Reliability International*, vol. 4, pp. 95-103.

Taguchi, G. (1986) *Introduction to Quality Engineering: Designing Quality into Products and Processes*, Asian Productivity Organisation, Tokyo, Japan.

Tsui, K,L. (1988) Strategies for planning experiments using orthogonal arrays and confounding tables, *Quality and Reliability International*, 4(2), 113-123.

Chapter 7   Inspection

Inspection is needed at various stages in a production operation: at the raw material input stage to ensure that only good material is fed to the production line, between production stages to ensure that only good products enter the next stage and at the end of the production line to prevent defective products from being passed to the customer. This chapter briefly discusses the role of inspection in quality control and reviews the application of artificial intelligence tools to inspection sampling plan selection and visual inspection.

7.1 Role of Inspection in Quality Control

The main objectives of an inspection system are:-

- to determine whether the manufactured product conforms to design specifications;
- to identify and separate non-conforming items from the production line;
- to compile a list of any deficiencies for short term and long term use.

As implied above, usually, the inspection process may involve the following activities [Juran *et al.*, 1962]:-

- interpretation of the given specifications;
- taking measurements of a product;
- making a comparison between the specifications and measured values;
- making a judgement on conformance;
- removal of non-conforming products;
- recording of the data obtained.

Kopardaker *et al.* [1993] have divided these inspection activities into 4 categories:-

(i) monitoring activities where the deviations from normal or acceptable norms of the process are tracked;
(ii) examining activities where defects in the produced items are searched for;
(iii) measuring activities where the status of an item is defined as good or bad;
(iv) patrolling activities which involve random sampling and spot checking.

The above authors have reported that among these activities, 'examining parts' is the kind of activity that both humans and machines are able to perform while machines are more suitable for 'monitoring' and 'measuring' activities and humans are more adept at 'patrolling' work. This observation leads to three types of inspection:-

(i) Manual inspection: humans perform all the inspection tasks. Several factors may affect the performance of a manual inspection system including the properties of the items produced, human visual ability, illumination, number of people involved (inspectors) and inspection policies of the company.

(ii) Hybrid inspection: some of the inspection activities are automated using tools such as static overlay or vision equipment. Because humans are still involved, the factors affecting the performance of manual inspection systems also apply here.

(iii) Automated inspection: all inspection activities are performed by machines. Computer vision, image processing and pattern recognition techniques have been developed to help to automate the inspection procedure. However, economic factors such as the cost of the inspection equipment and the cost of the peripherals required to operate with this equipment need to be considered when contemplating automating an inspection process.

Although a number of commercial inspection systems are available, their industrial use is not as widespread as expected due to the high costs of equipment and maintenance, problems with lighting systems and high-level programming requirements. However, the need to employ fully automated inspection is increasing due to rising labour costs and the necessity to guarantee 100% inspection. In the later sections of this chapter, automated visual inspection techniques are reviewed and the use of artificial intelligence tools for automated visual inspection is discussed.

A number of strategies are commonly employed to conduct inspection, such as spot checks, 100% inspection and sampling:-

Spot checking involves selecting a fixed percentage of a lot to be inspected, for example 10% of the lot. There is no scientific basis for this type of procedure. That is, spot checking does not employ statistical principles and does not provide information about the risk of incorrect decisions. This procedure is therefore more suited to quantity verification rather than quality verification and will not be discussed further here.

Sampling is widely used. It is based on using statistically determined samples. The inspected lot is either accepted or rejected according to the status of these samples. This method is suitable when 100% inspection is not possible.

100% inspection theoretically eliminates all non-conforming products, therefore stopping them from reaching the customer. It is very costly and impractical in most cases (especially in large production lots). However, there are some situations where 100% inspection is necessary, such as the inspection of products with critical safety requirements or those with high external quality failure cost. Recent technological developments have made it possible to devise fully automated inspection systems which are capable of performing 100% inspection. This point will be expanded later in this chapter.

7.1.1 Sampling

Sampling procedures have been developed to determine whether or not the quality of a product conforms to specifications using samples of the product instead of inspecting every item produced. The general procedure is shown in Figure 7.1. A 'lot' of items to be inspected is sent to an inspection station. Generally, random samples are selected and quality characteristics are measured and compared with some acceptance criteria.

If the acceptance criteria are met, the entire lot produced is accepted and sent to the related units or customer. Otherwise, the lot is rejected and a decision is made considering the following points:-

- can the rejected items be sold at a reduced price?
- is there a possibility for reworking the rejected lot?
- is there a need for 100% inspection?
- does the lot have to be returned to the supplier?

Sampling procedures have a statistical and mathematical background which makes them acceptable and implementable. Ready to use tables and procedures are given in quality control books which define the values and sampling parameters according to different sampling plans. See [Evans and Lindsay, 1989] for more details.

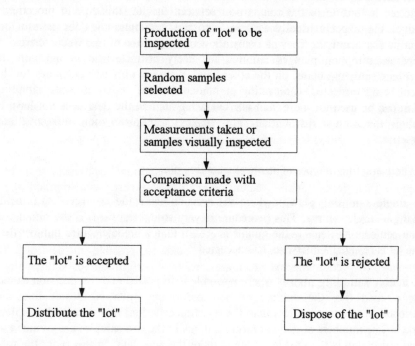

Figure 7.1 General sampling procedure

7.1.2 Sampling Plans

Sampling plans are procedures which provide an efficient sampling method for a particular inspection task. A good sampling plan should be able to protect the producer against the risk of rejecting acceptable lots (producer's risk) and protect the customer against the risk of accepting bad quality lots (customer's risk). It should also minimise the cost of sampling and provide information concerning the quality of the lot being inspected. Sample size and acceptance criteria are two important factors which may affect the design of such a plan. These parameters are determined according to the existing fraction defective (the proportion of non-conforming items in a lot), lot tolerance defective (the level of non-conformance that the customer is willing to tolerate a specified percentage of the time) and maximum allowed risk levels (producer's or customer's risk). There are various methods for determining these parameters. They differ according to the quality characteristic measured (attribute or variable).

In variable sampling plans, a measurement is taken on each item in the sample and a statistic, such as the mean or the variance, is calculated. This statistic is then compared against the allowed value or a critical value. This critical value is normally a function of the producer's or customer's risk. The lot is accepted or rejected according to the comparison between the lot statistic and the critical value. These plans usually assume a normal distribution for the measurable quality characteristic. They can only allow the inspection of one characteristic. A separate sampling plan is required for every characteristic in question. In attribute sampling plans, on the other hand, there are only two outcomes for the item being inspected: conforming or non-conforming. Each attribute sampling plan can be used for more than one characteristic and the data does not have to follow the normal distribution. This makes these plans more attractive and practical.

There are 3 main types of attribute sampling plan:-

A **single sampling plan** is where the decision about the product (lot) is made using a single sample. This procedure is very straightforward. If the number of non-conforming items in the sample is greater than a predetermined number then the lot is rejected. Otherwise, it is accepted.

A **double sampling plan** is where a sample is inspected and if a decision cannot be reached (i.e. the number of non-conforming units is greater than the acceptance criterion but less than the rejection criterion) then a second sample is taken. The numbers of non-conforming items in the two samples are summed up and a decision is then made on the basis of the new data. In this plan, the total number of units to be inspected decreases as "very good" or "very bad" quality products are disposed of using a smaller sample size while performing the first inspection. Only lots of "marginal quality" may require a second sample and therefore more inspection.

A **multiple sampling plan** is similar to a double sampling plan but uses more than 2 samples to accept or reject the lot in question. The sampling procedure continues until the cumulative results of inspection determine that an acceptance or rejection decision should be made. Note that the sample size is smaller in this case than for the other two sampling plans.

There are other specific inspection sampling plans that are applicable to different industrial situations. The reader is referred to the literature for more detail [Juran *et al.*, 1962; Wadsworth *et al.*, 1986; Evans and Lindsay, 1989].

7.1.3 Knowledge-Based Systems for Selecting Sampling Plans

The best plan for one product is not necessarily the most suited for another. Each product may require a different sampling plan for effective inspection When a large quantity of parts is involved, selecting an appropriate sampling plan which is a representative of the complete lot is an important but complex task. The producer's and customer's risks, costs and managerial policies are the main factors to be taken into consideration when adopting a sampling method. When any of these factors change, the sampling plan used may need to be updated. As with other quality tasks, intelligent knowledge-based systems have been used for the task of assisting quality planners in selecting an appropriate sampling plan, taking into account all the necessary factors just mentioned.

The general structure of an expert system called ASASP, developed for selecting sampling plans, is given in Figure 7.2. The knowledge base, inference engine and user interface are the basic components of the system.

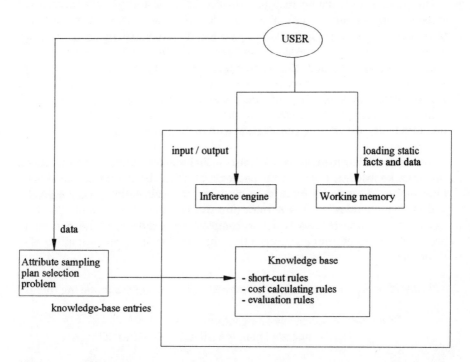

Figure 7.2 General structure of the ASASP expert system for sampling plan selection (adapted from Fard and Sabuncuoglu [1990])

The knowledge base is constructed by considering various qualitative and quantitative factors, such as:-

- purpose of sampling;
- allowed risk levels;
- fraction of lot defective;
- ease of application of the plan;
- cost of inspection, manufacturing and failure;
- sample size;
- average number of parts inspected;
- acceptable level of average output quality;
- lot size flexibility;
- flexibility of inspection time;
- managerial policies;
- acceptability of the plan to the producer.

Quantitative information may be provided using historical data, accounting records, engineering specifications or design requirements and customer tolerances. Qualitative information may be acquired from experts using their past experience. Once this knowledge has been extracted, the knowledge engineer needs to encode it into the knowledge base using a suitable knowledge representation technique. Note that some information is permanent and therefore can be stored in a database or static memory. Dynamic information which changes according to the properties of the process and user requirements can be provided by the user during consultation with the expert system.

The system shown in Figure 7.2 also includes rules extracted from general sampling knowledge. Fard and Sabuncuoglu [1990], who developed the system, have suggested 3 types of rules: short-cut rules, cost-calculating procedural rules and evaluation rules. Short-cut rules directly recommend a solution. In some cases, the selection process is not so complex. For example, if the number of acceptable non-conforming items is 0 then the selection of a single sampling plan is inevitable.

This information can be stored in the knowledge base using the following rule:-

> IF number of acceptable non-conforming items = 0
>   THEN suggested plan is a SINGLE SAMPLING PLAN

When the quality requirements are stringent, the related rules also become more complex. Where the cost of inspecting each piece is substantial, selection of a sampling plan would be beneficial despite its complexity. Wadsworth *et al.* [1986] have suggested a cost model which allows the calculation of the total cost for each sampling plan. This model is characterised by the following equation:-

$$TIC = a + bn + cf \qquad\qquad (7.1)$$

where:-

> TIC = total inspection cost;
> a = overhead cost;
> b = cost of selecting the sample (per unit);
> c = cost of inspection (per unit);
> n = maximum sample size;
> f = average sample size.

Once the cost has been calculated, the following rule will decide if multiple sampling is to be adopted:-

> IF no sampling plan is suggested
> AND total cost of multiple sampling < total cost of single sampling
> AND total cost of multiple sampling < total cost of double sampling
>   THEN suggest a MULTIPLE SAMPLING PLAN

It should be noted that this rule is general by nature. Some cases might require more detailed rules such as:-

> IF sample size < 100
> AND acceptance criteria < 2
> AND rejection criteria < 7
>   THEN suggest a MULTIPLE SAMPLING PLAN

This type of rule is to cater for a particular requirement. Expert systems should be capable of working with uncertain or incomplete knowledge. The systems developed should possess the ability to suggest the quality characteristics to be inspected and decide on the level of risks involved. They should be easily adaptable for the dynamic environment in which the product is manufactured.

7.2 Automated Visual Inspection

Human operators encounter difficulties in visual inspection tasks when the product size is small [Hou *et al.*, 1993] or the production rate is high [Pham and Alcock, 1996]. Also, for some critical products, it is required to perform 100% inspection which is a costly task when done manually. Automated inspection systems are required in these cases.

The basic components of an automated visual inspection system are given in Figure 7.3. These are a transporter, an inspection station, a processor and a sorter [Chin, 1988]. The transporter moves the object to be inspected into the inspection station where sensors (cameras), assisted by suitable lighting devices, capture images and send them to the processor for analysis. The processor then tells the sorter whether the object is good or defective. The processor is the heart of the system. It uses image processing and pattern recognition techniques on the captured image data to extract information about the status of the object. Detailed information on image processing and pattern recognition techniques for automated visual inspection can be found in [Batchelor *et al.*, 1985; Marshall and Martin, 1992; Bayro-Corrochano, 1993; Davies, 1990].

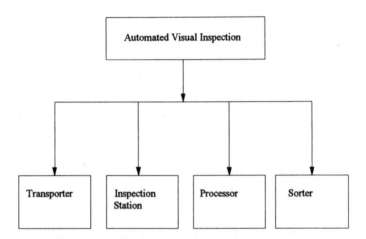

Figure 7.3 Components of an automated visual inspection system

The factors which need to be considered in an automated inspection system are reliability, versatility, frequency of false alarms, throughput, real-time operation, ease-of-use and cost. In the past decade, a number of successful industrial applications of automated visual inspection have been reported [Hattori *et al.*, 1992; Performance Imaging Inc., 1992]. Automated visual inspection systems have been developed for use in the electronic industry [Darwish and Jain, 1988; Glover, 1988; Sprague *et al.*, 1991], the food industry [Davenel *et al.*, 1988; Chan *et al.*, 1990; Wu and Rodd, 1991], the forestry industry [Pham and Alcock, 1996], the metal removal industry [Luk and Huynh, 1987], the textile industry [Longree, 1986] and the automobile industry [Pastorius, 1988; Elmaraghy and Bullis, 1989; Pham and Bayro-Corrochano, 1995; Pham *et al.*, 1995]. Some of the general advantages in the application of automated visual inspection are improved product quality, greater operator safety and comfort, higher productivities, increased protection for expensive production machinery and better use of materials [Batchelor *et al.*, 1985]. However, in spite of individual successes, it has not yet been possible to develop automated visual inspection systems that are sufficiently versatile and cost effective for all types of inspection applications. Existing systems still have to be developed specially for given tasks and they are also too costly, too difficult to implement and not reliable or fast enough for widespread industrial use. The next two sections will examine the possibility of enhancing automated visual inspection systems with knowledge-based and neural network techniques.

7.3 Knowledge-Based Systems for Automated Visual Inspection

Knowledge-based technology has been proposed for improving automated visual inspection systems. This technology is useful for high-level processing such as scene understanding, pattern analysis and interpretation [Tanaka and Sueda, 1988; Darwish and Jain, 1988; McKeown, 1987; Ntuen, *et al.*, 1989]. A typical knowledge-based system developed for visual inspection involves 4 modules (Figure 7.4). These are the knowledge module, processing module, data module and master program. The general structure shown in Figure 7.4 is intended to be effective for and adaptable to different applications. Having the knowledge module and master program separate creates a learning system which can be taught different models, rules and inspection algorithms.

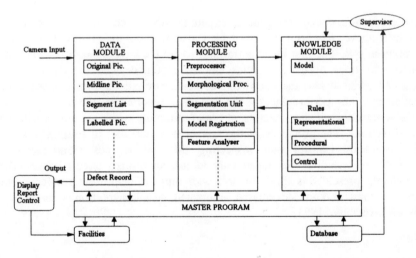

Figure 7.4 Knowledge-based system for visual inspection [Darwish and Jain, 1988].

The **knowledge module** is a long term memory where static knowledge is stored. This module includes rules and object models. Object models are special graphs representing geometric entities such as lines and angles. Each model is defined using several attributes such as labels, logical (semantic) information, geometric information, orientation, length, width and end points. Rules may be descriptive rules (describing the object to be inspected), procedural rules (relating to an inspection task) or control rules (coordinating the operation of the system). An example of a rule for an inspection task is given in [Ntuen *et al.*, 1989]:-

> IF compactness measure is less than 0.4
> AND average pixel mass is less than 26.00
> AND grey-level standard deviation is greater than 2.5
> AND histogram is bimodal
>   THEN product is defective

The **processing module** contains image processing and pattern recognition procedures which may require special hardware or firmware implementation.

The **data module** contains the image of the object under inspection and processed data generated from that image. This module acts as a short term memory to store the data needed or produced by the processing module.

The **master program** co-ordinates various system resources, synchronises the operation of the system and chooses the appropriate processing unit in the module to perform image analysis.

An inspection system for printed circuit boards, based on the components explained above, has been described by Darwish and Jain [1988]. It has two operation modes. The first mode is training where the system can be taught new models and rules or existing rules can be modified. The second mode is inspection where it uses the knowledge obtained through training to inspect the input image. The performance of the system is reported by the authors to be 100% in detecting all short circuits, cuts and minimum width violations with no false detections. The efficiency of the system is remarkable in finding missing and extraneous features and inspecting different regions simultaneously. Although the system is developed only to inspect printed circuit boards, the methodology presented is applicable to a wide range of inspection problems. For more examples of knowledge-based automated visual inspection systems, see [Bayro-Corrochano, 1993].

7.4 Neural Networks for Automated Visual Inspection

Neural networks can offer many benefits to automated inspection systems. These systems require speed, noise tolerance and independence from environmental changes during operation. Neural networks could effectively satisfy these requirements due to their parallelism and good generalisation capabilities under noisy conditions. This has attracted the interest of several researchers in applying neural networks for different vision tasks [Pawlicki, 1988; Beck *et al.*, 1989; Shi and Ward, 1989; Lee and Patterson, 1992; Pham and Bayro-Corrochano, 1994; Pham and Bayro-Corrochano, 1995; Pham *et al.*, 1995; Pham and Alcock, 1996].

Neural networks have been proposed for processing the image of an object to be inspected in terms of enhancement, segmentation, feature extraction and classification. A system using a neural network to classify defects on a product by examining features extracted from images of the product is schematically illustrated in Figure 7.5. The system forms the basis of an automated visual inspection machine for 100% inspection of automobile engine valve stem seals [Pham and Bayro-Corrochano, 1995; Pham *et al.*, 1995]. The machine, shown in Figure 7.6, can detect different types of defects on a seal.

There are three major tasks to be performed in order to implement a neural-network-based solution to an inspection problem. The first is to define the problem in such a form that the neural network can process it. This is a domain dependent task and requires extensive knowledge of the problem domain as well as neural network procedures and models. This task is interwoven with the other two tasks. The second task is to create good training and test data sets. The third is to find a suitable network topology. The latter two tasks will now be discussed in relation to the valve stem seal inspection problem. The performances of different types of neural network and rule-based defect classifiers developed for the seal inspection system will be presented.

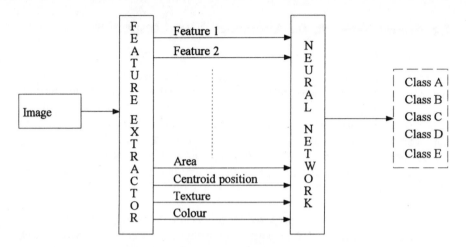

Figure 7.5 Neural network classification of defects

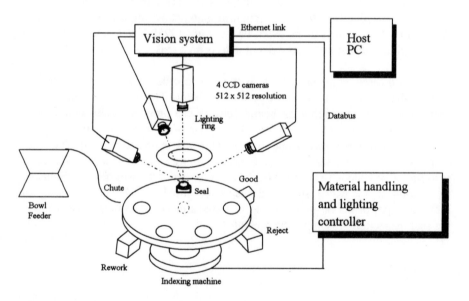

Figure 7.6 Valve stem seal inspection system

7.4.1 Training and Test Data Sets

A problem for an automatic inspection system is to find a good way of representing the essential features of the image of the object to be inspected in terms of feature vectors. Using raw image information requires a large amount of data which in turn gives input vectors with high dimensions. This increases the computational complexity and training time. Considering the noisy, redundant and irrelevant pixels in an image makes the problem more difficult to handle. It is therefore advisable to extract relevant features and use them instead of the raw data. Feature extraction methods are available which compress the image data into significant characteristics of the object under inspection. This reduces the dimension of the data and simplifies the information processing for real-time use. This simplification is also essential for neural-network-based inspection systems. If it is not carried out, then the network would typically be required to have hundreds of thousands of input neurons in order to handle the raw image data. Using features, on the other hand, allows the network to grasp the important properties of the object, which are stored in the input vector, without having to be concerned with other irrelevant details.

As seen in previous chapters, neural networks require two sets of data: one for learning the inspection task and one for testing the accuracy of the system. The training set is used by the neural network to extract classification rules. It is therefore very important that it contains good and representative examples. The test set is for assessing the generalisation ability, i.e. the performance of the classification rules on previously unseen data.

The training and test data sets for the seal inspection system comprise signature vectors and the defects they represent. The signature vectors are extracted from contour images of known defects. Each vector has 23 components, the first 20 of which record the frequencies of occurrence on a contour of the 20 different standard contour features. The features (straight line segments, curved segments, corners etc.) are detected using templates. Examples of these templates are shown in Figure 7.7. Figure 7.8 depicts the partial signatures consisting of the first 20 geometric feature components extracted for different defect types [Pham and Bayro-Corrochano, 1992]. The last 3 components of a feature vector are the values of the perimeter of the contour and the length and width of its minimum bounding rectangle (MBR). The MBR is the smallest rectangle that completely encloses the object represented by the contour and has its sides parallel to the edges of the image.

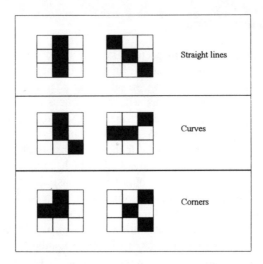

Figure 7.7 Examples of templates used for feature detection

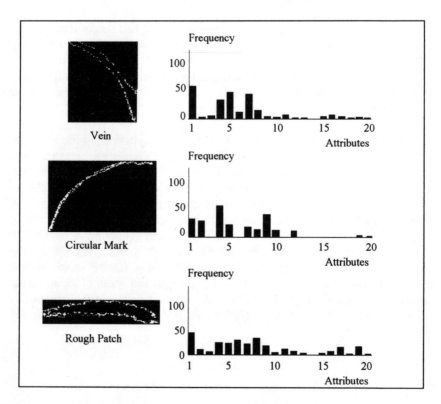

Figure 7.8 Defects and corresponding partial feature vectors

7.4.2 Neural Network Details

One of the neural networks used in the seal inspection system is a Multi-Layer Perceptron having 3 layers: an input layer with 23 neurons (corresponding to the 23 components of the feature vector), a hidden layer with 10 neurons and an output layer with 3 neurons (corresponding to the three types of defects to be identified: veins, circular marks and rough patches). The network is shown diagrammatically in Figure 7.9 and other parameters of the neural network are given in Figure 7.10.

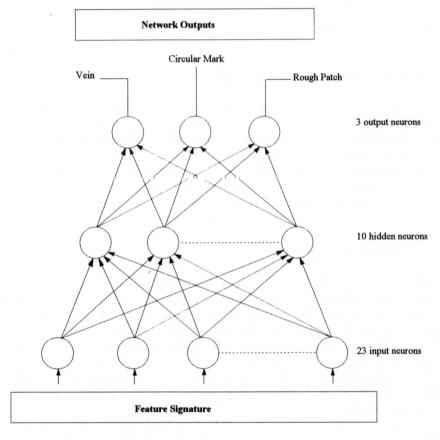

Figure 7.9 Topology of the MLP used

Number of input neurons	23
Number of hidden layers	1
Number of hidden neurons	10
Number of output neurons	3
Learning rate	0.7
Momentum coefficient	0.8
Maximum acceptable training error	0.001
Activation function	Sigmoid
Learning rule	Generalised delta rule
Number of iterations	130000

Figure 7.10 Parameters of the MLP used

7.4.3 Performance of the MLP Network

The network is trained for the task by presenting to it example feature vectors from the training data set in a random sequence and forcing its output to produce the correct defect class for each feature vector. A training session requires some 130,000 presentations of the complete training data set (of 180 feature vectors) before the output error decreases to below 0.001 (the output range being 0 to 1). The trained network is able to recognise correctly 93 out of 100 feature vectors in the test data set which it was not trained on.

7.4.4 Other Classifiers

The performance of the MLP has been compared against that of other classifiers on the seal inspection problem. Two of the new classifiers are also neural networks (LVQ and ART2) and the third is a rule-based classifier. The parameters of the neural networks are shown in Figures 7.11 and 7.12 respectively. The rules employed in the rule-based classifier are obtained manually by examination of the feature vectors and the corresponding defects [Bayro-Corrochano, 1993]. The percentage classification accuracies of the LVQ, ART-2 and rule-based classifiers on the same test data set are 85%, 62% and 72%, respectively.

Number of input neurons	23
Number of Kohonen neurons	30
Number of output units	3
Maximum acceptable training error	0.001
Activation function	Winner-takes-all
Learning rule	Minimum Euclidean distance
Number of iterations	2800

Figure 7.11 Parameters of the LVQ network used

Number of input neurons	23
Vigilance parameter	0.85
a	0.5
b	0.5
c	0.1
d	0.9
Number of iterations	1800

Figure 7.12 Parameters of the ART-2 Network used

7.4.5 Composite Systems

The best performance among the various classifiers is clearly that of the MLP. To obtain an improved performance, a composite inspection system is used that is made up of 3 individual MLP classifier modules (Figure 7.13). The system has a structure that is broadly similar to the structure of the composite system presented in Chapter 4. However, the decision making module, instead of being a rule-based module as in the system described in Chapter 4, is itself another neural network unit. The latter acts as a gating system to modulate and combine the outputs of the individual classifiers [Pham and Bayro-Corrochano, 1995]. The advantage of employing a neural network for this task is that there is no need to design explicit rules for combining the classifier outputs as those rules are implicitly developed by the neural network through training. As shown in Figure 7.14, the composite system has a classification accuracy of 96% which is higher than that obtained with any individual classifier.

Module Outputs

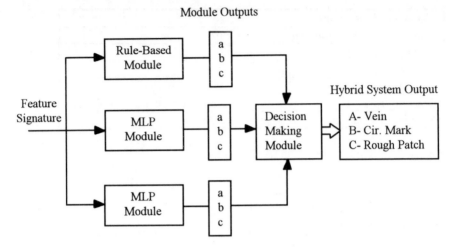

Figure 7.13 General structure of the composite inspection system

Classifier	Performance (%)
Rules	72
MLP	93
LVQ	85
ART-2	62
Composite system	96

Figure 7.14 Accuracies of the different classifiers

7.5 Discussion

The previous sections have illustrated the use of expert systems and neural networks to automated visual inspection. The aim of employing these artificial intelligence tools is to increase the reliability and speed of automated visual inspection systems. In the case of neural-network-based systems, the potential increase in speed will not be fully realised until truly parallel neural network hardware is available at affordable costs. Neural network solutions can then be extended to lower-level image processing tasks such as edge detection and filtering to improve the input to an inspection system [Pham and Bayro-Corrochano, 1992]. However, there is still a need for more research into utilising different neural network models in terms of hardware and software efficiencies.

The integration of neural networks and expert systems for automated inspection tasks has merits as the two tools possess complementary strengths. Although existing inspection systems have not yet exploited this technique, its feasibility has been demonstrated in Chapter 5.

7.6 Summary

This chapter has briefly explained the role of inspection in a quality system and discussed the use of knowledge-based expert systems and neural networks in inspection related tasks. Various applications of these artificial intelligence tools have been demonstrated, including applications to inspection sampling plan selection and automated visual inspection of valve stem seals.

References

Batchelor, D.G., Hill, D.A. and Hodgson, D.C. (1985) *Automated Visual Inspection*, IFS, Bedford and North Holland, Amsterdam.

Bayro-Corrochano, E.J. (1993) *Artificial Intelligence Techniques for Machine Vision*, Ph.D. thesis, University of Wales, College of Cardiff, UK.

Beck, H., McDonald, D. and Brazkovic, D. (1989) A self-training visual inspection system with a neural network classifier, *IEEE Int. Conf. on Neural Networks (INNS)*, Washington, vol. 1, 307-311.

Chan, J.P., Batchelor, B.G., Harris, I.P. and Perry, S.J. (1990) Intelligent Visual Inspection of Food Products, *Proc. SPIE OE / BOSTON '90: Machine Vision Systems Integration in Industry*, Conf. No 1386, Batchelor, B.G. and Waltz, F.M., SPIE, Washington, 171-179.

Chin, R.T. (1988) Automated visual inspection: 1981 to 1987, *Computer Vision, Graphics and Image Processing*, vol. 41, 346-381.

Darwish, A.M. and Jain, A.K. (1988) A rule-based approach for visual pattern inspection, *IEEE Trans. on Pattern Analysis and Machine Intelligence*, 10(1), 56-58.

Davenel, A., Guizard, C.H., Labarette, T. and Sevila, F. (1988) Automated detection of surface defects on fruit by using a vision system, *Journal of Agricultural Engineering Research*, vol. 41, 1-8.

Davies, E.R. (1990) *Machine Vision: Theory, Algorithms and Practicalities*, Academic Press, London.

ElMaraghy, H.A. and Bullis, D.J. (1989) Expert inspector for surface defects, *Computers in Industry*, vol. 11, 321-331.

Evans, J.R. and Lindsay, W.M. (1989) *The Management and Control of Quality*, West Publishing Company, St Paul, USA.

Fard, N.S. and Sabuncuoglu, I. (1990) An expert system for selecting attribute sampling plans, *Int. J. Computer Integrated Manufacturing*, 3(6), 364-372.

Glover, D.E. (1988) A hybrid fourier/electronic neurocomputer machine vision inspection system, *Robots 12, Vision '88, Conf. Proc.*, June 5-9, Detroit, MI, vol. 2, 8.77-8.103.

Hattori, T., Yamasaki, T., Nakada, M. and Kataoka, I. (1992) A practical architecture of high-speed image processor for automated visual inspection systems, *JAPAN/USA Symposium on Flexible Automation*, Kyoto, JAPAN, vol. 2, 1059-1066.

Hou, T.H, Lin, L. and Scott, P.D. (1993) A neural-network-based automated inspection with an application to surface mount devices, *Int. J. Production Research*, 31(5), 1171-1187.

Juran, J.M, Seder, L.A. and Gryna, F.M. (1962) *Quality Control Handbook*, McGraw Hill, New York.

Kopardaker, P., Mital, A. and Anad, S. (1993) Manual, hybrid and automated inspection: literature review and current research, *Integrated Manufacturing Systems*, 4(1), 18-29.

Lee, C.M. and Patterson, D.W. (1992) Deformation invariant recognition using a hybrid connectionist vision system, *Journal of System Engineering*, 2(4), 237-245.

Longree, M.A. (1986) System for automatic detection and recognition of defects in textile slivers, *Proc. of SPIE, Automated Optical Inspection*, Innsbruck, Austria, vol. 654.

Luk, F. and Huynh, V. (1987) A vision system for in-process surface quality assessment, *Vision '87 Proc.*, June 8-11, Detroit, MI, pp. 12.43 - 12.58.

Marshall, A.D. and Martin, R.R. (1992) *Computer Vision, Models and Inspection*, World Scientific, Singapore.

McKeown, D.M. (1987) The role of artificial intelligence in the inspection of remotely sensed data with geographic information systems, *IEEE Trans. on Geoscience and Remote Sensing*, 25(1), 330-348.

Ntuen, C.A, Park, E.H. and Kim, J.H. (1989) KIMS: a knowledge-based computer vision system for production line inspection, *Computers and Industrial Engineering*, 16(4), 491-508.

Pastorius, W.J., (1988) Vision applications in measurement, inspection and guidance related to large objects, *Robot 12, Vision '88 Conf. Proc.*, June 5-9, Detroit, MI, pp. 5.1-5.12.

Pawlicki, T. (1988) Recognising image invariants in a neural network architecture, *IEEE Int. Conf. on Neural Networks*, July 24-27, vol. 2, pp 135-142.

Performance Imaging Inc. (1992) *Image Analysis*, Library Manual Revision 2.1, January, Performance Vision Ltd, Solihull, West Midlands, UK.

Pham, D.T. and Alcock, R.J. (1996) Automatic detection of defects on birch wood boards, *Proc. I Mech E, Part E: J. of Process Engineering*, vol. 210 (in press).

Pham, D.T. and Bayro-Corrochano, E.J. (1992) Neural networks for noise filtering, edge detection and feature extraction, *J. of Systems Engineering*, 2(2), 111-122.

Pham, D.T. and Bayro-Corrochano, E.J. (1994) Neural classifiers for automated visual inspection, *Proc. I Mech E, Part D: J. of Automobile Engineering*, vol. 208, 83-89.

Pham, D.T. and Bayro-Corrochano, E.J. (1995) Neural networks for classifying surface defects on automotive valve stem seals, *Int. J. of Machine Tools and Manufacture*, 35(8), 1115-1124.

Pham, D.T., Jennings, N.R. and Ross, I. (1995) Intelligent visual inspection of valve stem seals, *Control Engineering Practice*, 3(9), 1237-1245.

Shi, P. and Ward, R.K. (1989) Using the perceptron to enhance binary noise images - Modelling and simulation on microcomputers, *Proc. 8th SCS, Western Conf.*, San Diego, California, pp. 119-123.

Sprague, A.P., Donahue, M.J. and Rokhlin, S.I. (1991) A method for automated inspection of printed circuit boards, *Computer Vision, Graphics and Image Processing*, 54(3), 401-415.

Tanaka, T. and Sueda, N. (1988) Knowledge acquisition in image processing expert system 'EXPLAIN', *Int. Workshop on Artificial Intelligence for Industrial Applications, IEEE Hitachi*, Japan, pp. 267-272.

Wadsworth, H.M., Stephens, K.S. and Godfrey, A.B. (1986), *Modern Methods for Quality Control and Improvement*, John Wiley and Sons, Singapore.

Wu, Q.M. and Rodd, M.G. (1991) Fast boundary extraction for industrial inspection, *Pattern Recognition Letters*, part 12, pp. 483-489.

Chapter 8   Condition Monitoring and Fault Diagnosis

This chapter discusses various issues and techniques related to machine and process condition monitoring and fault diagnosis. The use of expert systems and neural networks for condition monitoring and fault diagnosis is described.

8.1 Condition Monitoring

Condition monitoring is "the performance of periodic or continuous comparative measurements on parameters which are suspected of reflecting the condition of a component, subassembly or system with the object that, on analysis, the measurements may indicate the item's current condition and the future trend of its possible deterioration" [Davies, 1990]. Thus, *condition monitoring* can concern a range of equipment and techniques used to check a plant while it is in operation to give advance warning of deterioration and possible breakdown. Various parameters, such as vibration, temperature, speed and acoustic emission can be measured. These measurements can show the extent to which the plant has deteriorated. They can also highlight areas of poor system design. A good condition monitoring system should:-

- automatically monitor critical areas of the plant or equipment in real time;
- automatically capture, store, analyse and update information in the form of quantitative data for management reporting, adaptive control, maintenance, troubleshooting and diagnostic evaluation;
- activate alarm systems which can indicate the source of a potential failure, the sequence of events and the current operating status of the system under surveillance;
- provide an accurate database from which the long and short term failure patterns of the plant can be determined so that steps can be taken to protect initial plant installations against failure.

Performing the above activities would bring several technical, organisational and financial benefits to manufacturing [West, 1988; Henry, 1988; Williams *et al.*, 1994]. Technical advantages would include:-

- an increase in manufacturing system availability by minimising unscheduled equipment downtime and by reducing the repair time necessary for equipment restoration;
- a reduction in consequential damage by detecting incipient problems before they cause failure;
- an improvement in manufactured product quality and a reduction in the scrap rate;
- an improvement in safety by minimising the potential for destructive failure;
- an improvement in plant operation by reducing processing times, decreasing energy consumption and maintaining process accuracy and repeatability.

Organisational advantages would encompass:-

- a more efficient maintenance organisation;
- an increase in system availability by allowing the maintenance department to conduct more effective negotiations with equipment manufacturers concerning plant specifications, warranty periods, alarm systems, consequential damage and contract servicing;
- an improvement in safety, labour relations, productivity and quality control by predicting and preventing catastrophic failures.

Financial benefits are derived from minimising capital expenditure by achieving effective procurement of equipment, minimising operational costs by taking corrective actions to remove the underlying causes of failures, minimising the cost of repetitive failures attributed to the same cause and minimising maintenance expenditure by restructuring and reducing the size of the maintenance organisation.

8.1.1 Condition Monitoring Techniques

The main techniques for condition monitoring are:-

Visual Inspection: This is one of the most cost-effective methods of condition monitoring. It essentially consists of the visual inspection of machinery using either some form of optical assistance or the unaided eye. Automatic visual inspection techniques have already been discussed in Chapter 7. Readers may refer to Davies [1990] for a description of traditional visual inspection methods.

Performance Monitoring: This technique requires two preconditions to ensure successful application [Loynes, 1981]. First, the machinery to be monitored should be in normal operation and, second, measurements with respect to it must be recordable on a periodic or continuous basis. Providing these conditions are satisfied, any change in the normal operation of the system should be easily detected from changes in the measured data and a trend analysis may reveal the presence of a potential failure.

Vibration Monitoring: This method was developed mainly for rotating machinery [Nicholls, 1986; Watton, 1992; Williams *et al.*, 1994]. Machine vibrations are caused by the cyclic excitation forces set up in a system due to its operation. A machine produces vibration signals of different shapes when in a healthy state compared with a deteriorated state. Monitoring these signals can therefore provide information on the state of the machine.

Wear Debris Analysis: This method is useful when it is necessary to monitor machines having components in contact with a fluid such as hydraulic fluid or lubricant which are likely to be affected by progressive wear [Baur, 1982]. Examples of such components include bearings, gears, cylinders and seals [Watton, 1992].

8.1.2 Implementation Stages of Condition Monitoring

The implementation of a condition monitoring programme even on a small scale requires a good deal of effort. Experience is important to ensure that a correct approach is taken (especially during initial introduction). The usual stages of implementation include:-

(i) selection of personnel to plan and implement the project. The people who will be responsible need to be dedicated, competent, self-disciplined, trustworthy, logical, capable of clear and concise reporting and able to argue the case for the benefits of condition monitoring.

(ii) provision of initial awareness training. All members of the team formed in the first stage must be trained in the various techniques of condition monitoring. This ensures that the team is aware of alternative solutions to different condition monitoring problems. Knowledge-based systems can help in the training of the team members.

(iii) identification of pilot project plant, machinery or equipment. A rapid identification of critical equipment is normally achieved through the classification of all equipment into 4 different groups, namely:-

- *essential equipment* which must be on-line at all times to ensure production continuity and therefore requires frequent monitoring;
- *critical equipment* which may limit productivity, have high replacement costs and is known to have chronic maintenance problems;
- *important equipment* which is not critical but still requires monitoring to ensure acceptable plant performance in terms of quality and process efficiency;
- *other equipment* which is not critical but may be prone to failure.

(iv) selection of condition monitoring techniques. Once the plant and related equipment have been determined, an appropriate monitoring techniques need to be selected. The monitoring techniques must provide a long lead time between fault detection and ultimate equipment failure and be insensitive to external effects and random fluctuations in measurements. This implies that the task of selection requires knowledge and experience of the process, plant and equipment to be monitored.

(v) determination of the procedures, instrumentation and organisation for implementation. After a pilot project and appropriate condition monitoring techniques have been defined, a few critical components on a key machine, normally selected on the basis of previous failure knowledge, should be monitored for known and anticipated failure conditions. The data acquisition system adopted to capture the data required for condition monitoring should be assessed for its ability to provide accurate and easily interpreted information. This system should be integrated with the condition monitoring system so that the overall system is reliable, tolerant of false alarms, and has a good report generation capability and a low economic cost.

(vi) launching the condition monitoring programme. Successful launching requires that staff are available and trained in the techniques used and aware of the methods and procedures to be followed. All equipment must be in place and ready for use on prepared sampling points.

(vii) monitoring, evaluating and updating the programme. This final phase involves continually assessing the progress of the programme and making adjustments to it, if necessary, to ensure completely successful implementation.

8.1.3 Knowledge-Based Systems for Condition Monitoring

There have been a number of applications of knowledge-based expert systems to condition monitoring [Hoh, 1992; Martin, 1994]. As remarked earlier, condition monitoring tasks can require extensive human knowledge and experience which can be provided by computer through utilising intelligent knowledge processing techniques. An intelligent monitoring system should be able to recognise abnormalities ahead of catastrophic failures and initiate corrective actions without human intervention. To achieve this, the system should employ a decision making scheme which is able to interpret information from sensors, learn from the environment, adjust itself in response to knowledge gained during the learning process and decide on the appropriate corrective action. Knowledge-based systems incorporating the necessary domain knowledge can provide all of these functions.

Usually, in an on-line knowledge-based system for condition monitoring, the monitoring and processing of different signals are undertaken using general purpose software and manipulation of knowledge is conducted via a program written in a specialist artificial intelligence language such as LISP or PROLOG or implemented in an expert system shell. In general, knowledge-based systems are found to be most appropriate in situations where there are more faults than symptoms and where a large number of interrelated decisions must be taken to utilise a complex set of data [Hill and Smith, 1988]. These situations are becoming more common as companies try to capitalise on their sophisticated equipment by increasing the number of functional units to be monitored. Several reasons have been put forward to support knowledge-based applications in condition monitoring [Hill and Banes, 1988], for example:-

- the repeatability of a diagnosis given the same input data for frequent or rare events;
- the need to employ complex knowledge sources rather than a single source of knowledge;
- the desirable features of expert systems which are portable, always available and not subject to the frailties of human experts such as sickness and inability to be simultaneously present at two different places;
- the relative ease of updating a knowledge-based system as new experience is gained. (Note, however, that there are not many systems which can effectively learn new rules as more data becomes available);
- the provision of safeguarded knowledge which is not lost when the expert leaves employment;
- the higher cost effectiveness of knowledge-based systems compared to human experts.

A number of knowledge-based systems have been developed for a variety of condition monitoring tasks[1]. Martin [1994] has reviewed systems that monitor the state of machine tools. Szafarczyk [1994] has compiled descriptions of condition monitoring applications in metal cutting, metal forming, automatic

---

[1] Condition monitoring and fault diagnosis are often implemented in the same system. The examples cited here are of knowledge-based systems that perform both tasks.

assembly and other manufacturing operations. Watton [1992] has extensively discussed the condition monitoring of fluid power systems and given details of expert systems for monitoring leakages in hydraulic motors and cylinders. The systems described incorporate diagnostic rules embedded in a simple but effective rule-based shell. One of the systems also contains rules for checking the integrity of sensors as part of its monitoring procedure. The author has compared this expert system approach and the conventional programming approach. The advantages of the former include requiring a shorter and simpler program and providing a better user interface with explanation facilities. Finally, Bridson *et al.* [1990] have presented an expert-system-based condition monitoring system for use in helicopters. The system monitors vibration signals and discriminates between normal and abnormal conditions. It employs an evaluation function to decide the severity of a condition and the rate of deterioration. The system interfaces with another expert system which is ground-based and which provides maintenance advice.

8.1.4 Neural Networks for Condition Monitoring

As mentioned before, by monitoring the condition of a machine, failures can be anticipated and maintenance can be scheduled to minimise disruption. Determining the condition of a machine in operation is often difficult even for trained experts. Researchers have investigated the use of neural networks to automate this task. For example, Wasserman *et al.* [1991] have developed a neural-network-based system for monitoring the condition of roller bearings. The system is designed for vibration data although other measurable variables such as pressure, temperature, oil analysis and acoustic data can also be input to the neural network. The reason for monitoring roller bearings in large rotating machinery is that failure can be expensive and dangerous. Vibration signatures are used to define the condition of the bearings. However, due to noise and reverberation, signals can be corrupted and, without the neural network, the determination of the condition of a machine would require expert human judgement. Classical pattern recognition procedures are reported to have limited success with such noisy and distorted signals [Wasserman *et al.*, 1991].

In the system developed by Wasserman *et al.*, vibration data from accelerometers mounted on the housings of 7 bearings in 2 different machines is recorded on a magnetic tape. Data samples are taken when the bearings are determined to be in good and bad conditions. In total, 280 signatures, each having 8 features, are derived using time, frequency and spectral domain techniques. The signals are normalised before being input to the network. After training, the neural network gives the same result as a bearing vibration expert with a very high confidence level. The authors have reported similar results for another application involving the monitoring of crack development in a shaft.

Dornfeld [1990] has presented a real-time cutting tool wear monitoring system which employs neural networks for sensor fusion and decision making. The system is able to recognise abnormalities and suggest corrective actions. Several sensors are used to measure cutting force and spindle motor current. A Multi-Layer Perceptron is trained on these signals. Training is performed off-line but the system is used on-line. Significant improvement in tool wear detection is achieved by fusing information from multiple sensors. The author has reported that:-

- neural networks can effectively detect tool wear but their parameters have to be selected carefully. Small networks perform better than larger ones on this task;
- the neural network tool wear monitoring system is insensitive to changes in cutting conditions and can be dynamically adjusted to operate within a wide range of conditions even if it is trained only on one particular condition;
- the system can operate on-line in real time.

Similar results have also been obtained by Monostori [1993] using neural networks for the same task.

8.2 Fault Diagnosis

The aim of fault diagnosis is to discover the cause or causes of a set of symptoms. A fault is a discrepancy between the actual structure of a device and its design. When the device is faulty, its behaviour is different from its expected behaviour. A symptom is any observable feature that is inconsistent with the expected behaviour of the device [Genesereth, 1984 and 1985].

A general structure for a diagnostic process is given in Figure 8.1. This shows 4 stages:-

- **fault detection** which is the recognition of a fault in a system. This may involve the observation of a defect in a product. If any fault is present then the observation should indicate a deviation of the product from its specifications. The result of this analysis will confirm some symptoms or reject others;

- **fault isolation** which constrains a fault to a sufficiently small sub-region or module of the system. At this stage, several hypotheses need to be developed about the causes of the existing malfunction. This requires extensive knowledge of how the system operates and what the relationship between the causes and the symptoms is;

- **fault identification** which points out the reasons for the failure. At this stage, the most likely hypothesis is chosen to reach a diagnosis. If no hypotheses are found to confirm a fault then it might be necessary to carry out more tests to create new hypotheses. This process is repeated until an acceptable diagnosis is reached;

- **fault correction** which suggests appropriate corrective actions to rectify the faults.

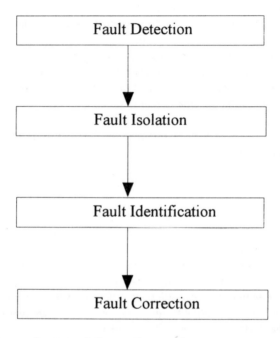

Figure 8.1 Stages of a general diagnostic procedure

Fault diagnosis is thus a process of hypothesis generation and verification. The principal aim is to obtain sufficient information to diagnose the faults. However, creating and verifying hypotheses is a non-trivial task. It requires experience and knowledge about the given problem domain. The task becomes even more difficult when experiencing faults previously not encountered. Many approaches have been suggested for automating the process of fault diagnosis, each having their strengths and weaknesses.

The main automatic fault diagnosis techniques described in the literature include:-

- **decision trees** (fault trees) which have been developed to generate a minimum set of tests and measurements in order to locate a faulty unit based on the theory of Boolean algebra [Davis, 1984]. The technique is suitable for diagnosing "hard" faults in binary elements of a system [Williams *et al.*, 1994]. The construction of fault trees requires a great deal of effort. Also, this method is not cost effective when applied to a fault free device;

- **fault dictionaries** which are systems storing the signatures or symptoms associated with different types of faults [Williams *et al.*, 1994]. A deficiency with fault dictionaries is that they require prior knowledge of all the faults to be encountered and all the symptoms corresponding to those faults. This prior knowledge is not always available. Fault dictionaries have been applied to detect single and multiple related faults [Ishida, 1985]. In the latter case, the interrelationship between faults must also be known;

- **model-based techniques** which use models of the behaviour, functionality or structure of the device to be diagnosed in addition to a consideration of its structure [Bandekar, 1989; Stells, 1989]. There are considerable advantages in using models. They help predict the behaviour of the actual system and give an insight into the designed function. It is therefore easy to observe differences between the actual and predicted outputs. The most important feature of model-based fault diagnosis systems is that they can handle new and unanticipated faults. This is not possible with simple rule-based techniques which are fragile at the boundary of the stored expertise. The required information for implementing a model-based fault diagnosis system is readily available from the design process. However, the main problem with the model-based approach is to find a correct model. The diagnostic efficiency is only as good as the diagnostic model. Because of the complexity involved, these systems have not been widely applied in practice.

The diagnostic process is likely to involve the analysis of structural and functional descriptions of the product or system in question. There is a need to have adequate experience and process knowledge, inferencing capabilities and problem solving abilities in uncertain conditions. Artificial intelligence techniques can naturally provide a solution to this need. In particular, due to their ability to solve problems requiring human expertise, knowledge-based tools can be used to implement intelligent diagnostic systems to assist diagnosticians to correct failures and secure the quality levels required in modern manufacturing.

8.2.1 Knowledge-Based Systems for Diagnosis

Knowledge-based techniques are inherently designed to deal with incomplete and vague, behavioural and functional information. Knowledge-based systems have been used for diagnosis of several processes and systems including manufacturing processes and machine tools, VLSI chips and computer systems, turbine generators, steam boilers, electronic circuits, power networks, communication networks and chemical plants.

There are two functions to be performed by a knowledge-based diagnostic system. They are (i) analysing the observed symptoms to locate the causes of failure, and (ii) suggesting possible corrective actions. Note that these functions require extensive experience and knowledge because there may be more than one fault causing the symptoms and the relationship between the causes and symptoms is often ambiguous. The basic knowledge required is the set of malfunctions and the relationship between them and the observations. This knowledge may be obtained using the structural, behavioural, functional and compiled information available. Milne [1987] has suggested a hierarchical structure of diagnostic knowledge sources and the different types of systems embodying them (see Figure 8.2).

Functionally, a knowledge-based diagnostic system should [Spur and Specht, 1992]:-

- provide knowledge about causal relationships when faults arise so as to enable even those operators who are not well acquainted with the system or process being diagnosed to localise the causes of faults;
- give information on the consequences of faults so that the severity of the effect of a given fault on the rest of the production process can be estimated;
- direct the user to eliminate the faults (i.e. repair instructions should be available to him in a suitable form);
- have capabilities to help operators with no previous knowledge and experience in computing;
- offer facilities to the operator to maintain the system after a short period of training so that engineers with specialised knowledge can be relieved of these tasks.

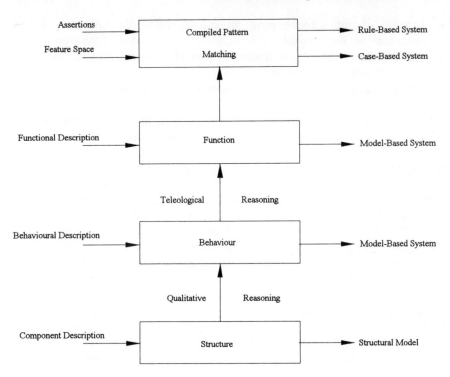

Figure 8.2 Hierarchical structure of diagnostic knowledge [Milne, 1987]

The construction of a diagnostic knowledge base is a difficult task. This is because the knowledge base contains the knowledge required to diagnose a process or a product and the construction must rely on the experience of the domain expert and his ability to express this knowledge clearly. If the expert is not good at articulating his knowledge in a systematic way, it is not possible to represent and formalise it, which reduces the efficiency of the system. There are several methods for acquiring knowledge [Cullen and Bryman, 1988]. When applied correctly, these should facilitate the development of the knowledge base.

A diagnostic knowledge base for a manufacturing system may contain the following components [Grob and Pfeifer, 1990]:-

- a model of the manufacturing machines in the system to define the characteristics that are important and the kind of faults that may occur;
- a model of the product to identify various likely defects during manufacturing;

- a model of the manufacturing process to define the sequence of the manufacturing steps together with the relevant components of the machine;
- a description of causal and functional dependencies determining the cause and effect relationship of a fault;
- a description of the remedies or corrective actions to be taken.

As mentioned previously, a benefit of knowledge-based systems is their ability to allow the user to add or delete information without adversely affecting the existing system. In practice, new situations often arise and so knowledge needs to be updated and the new knowledge then has to be validated. Validation is a difficult issue as an expert could be highly subjective in presenting his knowledge. Also, in most cases, the rules that encapsulate diagnostic expertise represent only the relations between observed symptoms and their causes without any explicit information about functional dependencies. This is the reason why, with simple rule-based systems, unfamiliar symptoms cannot be handled at all. There have been attempts to address this deficiency. For example, Yudkin [1988] and Havlicsek [1989] have proposed a structural based approach where a large rule base is restructured in hierarchical levels of diagnosis in order to provide a more flexible mechanism for hypothesis selection. As already seen in Figure 8.2, the knowledge about the device or product to be diagnosed can be organised as a hierarchical structure. This process of structuring diagnostic knowledge also simplifies the development and maintenance of the knowledge-based system, allowing easy change with small local effects. At the same time, the rule execution speed is increased since only relevant rules are considered. Peng and Ragia [1987] have developed another approach described as a probability-based approach. This approach uses a merit function to find the most probable hypotheses. The failure diagnostic model of Ishida [1985] involves a set of units which are faulty, a set of measurements to detect faulty units and a binary relation between these two sets. All possible combinations of fault patterns are considered under a permissible number of fault categories.

Alternatively, a small number of model-based systems have also been employed for diagnosis [Dvorak and Kuipers, 1991; Bau and Brezillon, 1992; Lackinger and Nejdl, 1993]. As already mentioned, in these systems, the behaviour and functionality of the components are modelled usually in addition to their structures. The behaviour of the system can be defined with a set of rules describing the relationship between the inputs and outputs of each component. It is the model itself that determines the performance of the system and construction of a good model is a very complex task.

Each knowledge engineering approach has its advantages and disadvantages. Integrating several approaches in a hybrid environment should help to obviate the weaknesses of each approach. There have been several attempts to integrate different diagnostic techniques to form an efficient diagnostic knowledge-based system [El-Ayeb and Finance, 1988; Lee, 1990; Graham *et al.*, 1993]. An integrated diagnostic system can have a hierarchical structure which allows recursive navigation through the hierarchical levels to improve the knowledge or rule set at each level. If the system fails to reach a diagnosis, it switches to the next level, going deeper into the model using simulation-based expertise. The benefit of hybrid systems incorporating structural and functional knowledge is that the combination of these two types of knowledge provides closure and completeness. Instead of requiring a large number of rules, possible faults can be derived from the description of the system by applying simple inference mechanisms.

Experimental systems utilising the above approaches have been developed [Fink and Lusth, 1987; Mitchell, 1989; Pflueger, 1989; Mauer, 1990]. WIND [Wilkinson, 1985] is an expert system for VLSI test system diagnosis and is a good example of a hierarchically structured rule-based system. It illustrates how the hierarchical structure reduces repair mean time. PROD [Odryna and Strojwas, 1985] is a diagnostic expert system that analyses the joint probability density function of measured IC parameters to determine the sources of faults in chips. EDARS [Ghafoor and Kershaw, 1989] is a general hypothetical expert diagnosis and repair system which can be used to diagnose and repair any system being monitored. DIVA [David and Krivine, 1989] is an expert system for vibration-based monitoring of turbine generators. The aim of the system is to help plant operators to interpret the evolution of the vibrations and diagnose the developing faults.

HYPOSS [Yang and Chang, 1989] is a hybrid diagnostic expert system which supports decision making in a nuclear plant.

DIAD-Kid/Boiler [Calandranis *et al.*, 1990] is an on-line performance monitoring and diagnostic system for the process industry. The system has two main modules, namely, the front interface module (FIM) and the reasoning module (RM). FIM is developed in an object-oriented environment called ACTOR. RM is developed using an expert system shell called NEXPERT. TACKER [Church, 1990] is an expert system for diagnosis and repair of electronic systems. This system is a general system which allows a range of diagnostic applications and has been successfully employed at British Telecom.

RBEST [Braunwalder and Zaba, 1990] is an expert system which determines the cause of failure in disk drives during the final 24-hour environmental stress test at the end of the manufacturing process. The system can successfully diagnose 98% of the faulty drives. KANT [Ronchi *et al.*, 1990] is an expert system which has been designed to diagnose communication boards of computers. 85% of the faults can be detected and corrected automatically and the remaining 15% dealt with successfully by the system with the interaction of a specialist.

ACDC [Fanni *et al.*, 1993] is another hybrid expert system capable of diagnosing faults on analogue circuits under dynamic conditions. The system works by triggering a causal propagation mechanism. Its diagnostic module analyses the circuit's behaviour and compares it against the desired behaviour to detect the candidate failure.

8.2.2 Neural Networks for Diagnosis

Fault detection and diagnosis may be considered a pattern recognition and classification problem. It has been stated that neural networks perform the classification and recognition of patterns better than other conventional methods [Lippmann, 1989]. Neural networks have an associative memory which allows them to develop an associative diagnostic ability with respect to faults that may occur in a process or product. They have been used in diagnostic tasks as they can handle many complex symptom-fault relationships and can cope with large volumes of data effectively. This is due to the fact that neural networks can be taught multi-dimensional mappings. This capability becomes more useful as the complexity of the mappings increases.

Neural networks are also attractive due to their self-adapting capability [Hassoun, 1995]. This allows a diagnostic process to be learnt quickly to suit changing conditions and degraded equipment. Above all, neural networks can readily be employed for on-line real-time diagnosis as they have a very short response time. The noise filtering ability of neural networks is an added feature which makes them appropriate for use in industrial environments characterised by noisy data. It is thus not surprising that neural networks have been applied for fully automatic diagnosis of faults [Watanabe *et al.*, 1989; Davies *et al.*, 1994].

As mentioned before, one of the main problems in developing an automatic diagnostic system is to acquire the experience of the expert diagnostician and represent it in a systematic way. In contrast to traditional rule-based systems, neural networks can make it easier to extract expert knowledge from raw data, overcoming the knowledge acquisition bottleneck by utilising the raw data directly from existing databases through a training procedure. This knowledge is represented using a parallel distributed structure. During knowledge acquisition, the values of predictor features (symptoms) serve as inputs to the network and output neurons are assigned values indicating the classes of faults associated with those predictor features (see Figure 8.3). Information about faults can be built into neural networks by training them on a set of data such as the steady-state process variables for normal conditions and those for identified faulty conditions.

This data can be provided by the past history of the process or machine being diagnosed. Training a neural network, as usual, involves adjusting the connection weights so that the network outputs become close to the desired outputs or expected goals. Weight changing is repeated until the network has correctly learnt the essential features of the input data. Applying a neural network in this way produces a trainable classifier, the goal of which is to learn complex non-linear dependencies between patterns (feature vectors) and their type (class) using a training set.

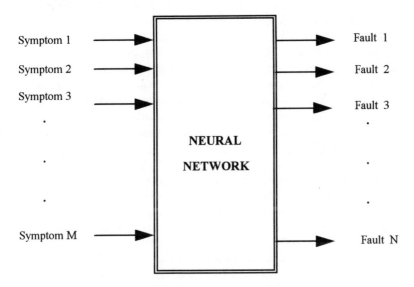

Figure 8.3 A general diagnostic neural network

Once a network has been trained, it can be used to diagnose faults on-line. When presented with an input value, the network generates an appropriate output value using its associative memory. The validity of this output totally depends on the network's mapping of the underlying input/output (symptom/fault) relationships. This clearly indicates the importance of the data set used to train the network. It must first cover the full problem domain and, second, contain sufficient information to show cause and effect relationships unambiguously. Note, however, that the performance of a network is not only affected by the training set but also by the method of encoding the input data [Cherkassky and Lari-Najafi, 1992].

A number of neural-network-based diagnostic systems have been described in the literature. VerDuin [1990] has analysed a system developed to diagnose machine faults. Watanabe *et al.* [1989] have described a diagnostic system using neural networks for a chemical process. They have proposed a two-stage neural network architecture using a Multi-Layer Perceptron in each stage. In the first stage, the task of the network is to discriminate among the causes of the faults and, in the second stage, the aim is to estimate the levels of a fault already identified in the first stage. Several networks are employed in the second stage, one for each fault. This is to allow new knowledge to be more easily incorporated. When new knowledge about the level of a certain fault becomes available, only the network corresponding to that fault has to be retrained.

Sobajic *et al.* [1989] have applied neural networks to diagnose and monitor power system operating conditions on-line. They have used unsupervised learning processes characterised by stable cluster formation within a small number of training epochs. They have found that neural-network-based monitoring and diagnosis is ideal for real-time implementation due to the extremely fast response and adaptation capabilities of a neural network. Marko *et al.* [1989] have also reached the same conclusion and found that pattern classification of faults is possible in real time using neural networks. Their system is able to diagnose 16 different faults in an automobile engine in real time with a high degree of accuracy.

Venkatasubramanian and Chan [1989] have utilised neural networks to solve fault diagnosis problems in a fluidised catalytic cracking unit (FCCU). The neural networks are integrated into an expert system called CATEX. This system is able to cope with 18 symptoms and 13 faults. The authors have tried different Multi-Layer Perceptrons and the best diagnostic performance achieved is 98.18%. Dietz *et al.* [1989] have developed a neural-network-based jet and rocket engine fault diagnosis system for real-time use. Their system successfully identifies the type of faults and determines the severity and duration of the faulty condition. Calabrese *et al.* [1991] have employed neural networks to generate the most likely cause hypotheses for a particular failure. Symptoms and causes are given to the network as inputs and outputs respectively.

Using Neural Networks to Diagnose a Coolant System: This section describes a neural-network-based fault diagnosis system developed for the coolant system in a vertical milling machine [Marzi and Martin, 1991; Martin and Marzi, 1993]. The coolant system is an important component in a metal cutting machine and its failure can lead to damage to the cutting tool and in some cases to the machine itself. Monitoring of a coolant system is usually performed by checking the signal from a control valve or pressure sensors. The main aim of the diagnostic system developed by Martin and Marzi is to relate the transient response of the system to the state of operation of the machine by using information collected from fault simulation.

Training Data: Diagnostic information is stored as transient pressure response patterns and associated system faults in a fault dictionary. Each pattern consists of 100 points. There are 3 possible faults: blocking of the filter, partial opening of the control valve and failure of the relief valve to function correctly. Each fault is described at 4 different levels of severity. The training data contains examples of each fault level. Data has also been collected at different time intervals for a cooling system in good working order to represent a range of healthy behaviours.

Neural Network Structure: The networks used are Multi-Layer Perceptrons with 4 layers. The input layer consists of 100 neurons. There are two hidden layers having 30 and 10 neurons respectively. These numbers have been obtained through experimentation. The output layer has 8 output neurons for encoding 4 categories of outputs ((a) healthy, fault 1, fault 2, fault 3, or (b) fault level 1, fault level 2, fault level 3, fault level 4).

Results: A two-stage diagnosis procedure is employed. In the first stage, a neural network is employed that has learnt all the available faults and their levels. The main aim of this stage is to discriminate between healthy and faulty states and to identify the fault type, if the system is deemed to be faulty. In the second stage, three different neural networks are used whose tasks are to define the level of the fault already identified in the previous stage (see Figures 8.4 and 8.5). It is interesting to note that in this application, the momentum coefficient does not help in the learning process and therefore is set to 0. The trained system has been tested on 56 patterns. The results obtained have shown that the network is able to recognise 100% of the test patterns in the first stage and correctly diagnose the levels of faults 95% of the time in the second stage.

8.3 Discussion

A method of fault diagnosis must characterise how a set of abnormal symptoms are related to a particular fault. This requires storing, retrieving and processing a large amount of knowledge, which is why artificial intelligence techniques can be useful for this task. In particular, knowledge-based systems have proved capable of solving difficult diagnostic problems. In these systems, the diagnostic knowledge takes the form of rules associating symptoms and faults. Sometimes, a qualitative or quantitative model of the process is developed to accompany the knowledge-based system. This enables the cause of the observed process abnormalities to be deduced through reasoning and simulation with the model. However, the main problem in formulating a diagnostic system is the difficulty of creating a model or a rule set which can cope reliably with dynamic environments.

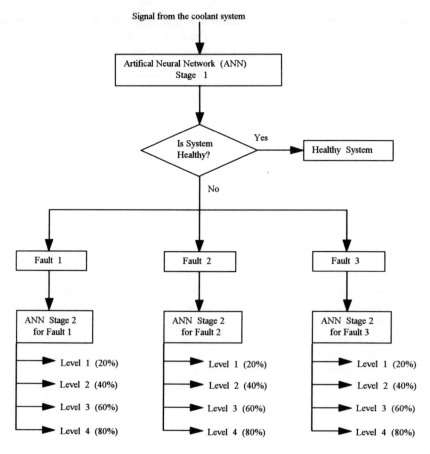

Figure 8.4 Neural-net-based coolant system diagnosis [Martin and Marzi, 1993]

Neural Network	State of coolant system	Fault level				Target value
Stage 1	Healthy	-	-	-	-	11001100
	Fault 1	20%	40%	60%	80%	11110000
	Fault 2	20%	40%	60%	80%	00110011
	Fault 3	20%	40%	60%	80%	00001111
Stage 2	Faults 1, 2, 3	20%				11001100
	Faults 1, 2, 3		40%			00110011
	Faults 1, 2, 3			60%		11110000
	Faults 1, 2, 3				80%	00001111

Figure 8.5 Neural network outputs for different states of the coolant system

This has led to the introduction of adaptable approaches to diagnosis. Adaptable systems employ neural networks to learn the relationship between symptoms and faults using examples of the actual data produced by the process to be diagnosed. The data collected from the process replaces the expert knowledge and explicit system model. Adaptable approaches are particularly suitable when expert diagnostic knowledge and prior models of the symptom/fault relationships are not available or easy to construct.

Although artificial-intelligence-based approaches outperform other traditional approaches, they are not without their problems [Kramer and Leonard, 1990]. For example, creating a data set representing the complete range of faults in a system is not a trivial task. In the absence of real data, the only way to obtain data is via conducting simulations of the process. Moreover, if the sensors in a system are faulty, they can produce corrupted data which would cause inconsistencies in the training data set. When new faults occur then the system needs to be retrained. Otherwise, cases falling in regions outside the scope of the training data could be misclassified. As neural networks do not have explanation capabilities to bring these cases to the attention of the operator or diagnostician, the integration of knowledge-based and neural network techniques would produce more stable and robust diagnosis systems. The previous section has given an example of a system combining neural network and knowledge-based techniques [Venkatasubramanian and Chan, 1989]. A full discussion of the benefits of integrating these techniques can be found in Chapter 5.

8.4 Summary

This chapter has discussed the monitoring and diagnosis of machines and processes. Condition monitoring and diagnostic procedures, techniques, and application steps have been outlined. The use of knowledge-based systems and neural networks in condition monitoring and diagnosis has been presented.

References

Bandekar, V.R. (1989) Causal models for diagnostic reasoning, *Artificial Intelligence*, vol. 29, pp. 3-32.

Bau, O.Y. and Brezillon, P.J. (1992) Model based diagnosis of power-station control system, *IEEE Expert*, February, pp. 36-40.

Baur, P.S. (1982) Ferrography: Machinery-wear analysis with a predictable future, *Power*, 126(10), 114-117.

Braunwalder, K. and Zaba, S. (1990) RBEST: An expert system for disk failure diagnosis during manufacturing, *Practical Experience in Building Expert Systems*, Branur, M. (ed.), Wiley, Chichester, UK, pp. 125-146.

Bridson, D.W., Atkinson, R.M. and Woolons, D.J. (1990) *The use of expert systems in advanced condition monitoring*, Technical Report, School of Engineering, University of Exeter, UK.

Calabrese, G., Gnerre, E. and Fratesi, E. (1991) An expert system for quality assurance based on neural networks, *Parallel Architectures and Neural Networks*, *4th Int. Workshop*, International Institute for Advanced Scientific Studies, Salerno, Italy, May, pp. 296-300.

Calandranis, J., Stephanopoulos, G. and Nunokawa, S. (1990) DIAD-Kit/Boiler: On-line performance monitoring and diagnosis, *Chemical Engineering Progress*, January, pp. 60-68.

Cherkassky, V. and Lari-Najafi, H. (1992) Data representation for diagnostic neural networks, *IEEE Expert*, October, pp. 43-53.

Church, C. (1990) TRACKER: Lessons from a first expert system, *Practical Experience in Building Expert Systems*, Branur, M. (ed.), Wiley, Chichester, UK, pp. 50-73.

Cullen, J. and Bryman, A. (1988) The knowledge acquisition bottleneck: Time for reassessment, *Expert Systems*, 5(3), 216-225.

David, J. and Krivine, J. (1989) Augmenting experience-based diagnosis with causal reasoning, *Applied Artificial Intelligence*, vol. 3, pp. 239-248.

Davies, A. (ed.), (1990) *Management Guide to Condition Monitoring in Manufacturing*, Institution of Production Engineers, London.

Davies, A., Thomas, P.V. and Shaw, M.W. (1994) The initialisation of artificial intelligence to achieve availability improvement in automated manufacture, *Int. J. Production Economics*, 37, 259-274.

Davis, R. (1984) Diagnostic reasoning based on structure and behaviour, *Artificial Intelligence*, vol. 24, pp. 347-410.

Dietz, W.E., Kiech, E.L. and Ali, M. (1989) Jet and rocket engine fault diagnosis in real time, *Journal of Neural Computing*, 1(1), pp. 5-18.

Dornfeld, D.A. (1990) Neural network sensor fusion for tool condition monitoring, *Annals of the CIRP*, 39(1), 101-105.

Dvorak, D. and Kuipers, B. (1991) Process monitoring and diagnosis: A model based approach, *IEEE Expert*, June, pp. 67-74.

El-Ayeb, B. and Finance, J.P. (1988) On co-operation between deep and shallow reasoning SIDI: An expert system for troubleshooting diagnosis in large industrial plants, *Artificial Intelligence in Engineering: Diagnosis and Learning*, Gero, J.S. (ed.) Elsevier and Computational Mechanics Publications, Southampton, pp. 95-118.

Fanni, A., Diana, P., Giua, A. and Parezzani, M. (1993) Qualitative dynamic diagnosis of circuits, *AI EDAM*, 7(1), pp. 53-64.

Fink, P. and Lusth, C. (1987) Expert systems and diagnostic expertise in the mechanical and electrical domain, *IEEE Transactions on Systems, Man, and Cybernetics*, SMC-17, Vol. 3, pp. 411-436.

Genesereth, M.R. (1984/5) The use of design descriptions in automated diagnosis, *Artificial Intelligence*, vol. 24/25, pp. 1-5.

Ghafoor, A. and Kershaw, R.S. (1989) A design methodology for expert systems for diagnosis and repair, *Proc. of the IEEE 8th Annual Int. Phoenix Conference on Computers and Communication*, pp. 550-554.

Graham, J.H., Guan, J. and Alexander, S.M. (1993) A hybrid diagnostic system with learning capabilities, *Engineering Applications of Artificial Intelligence*, 6(1), 21-28.

Grob, R. and Pfeifer, T. (1990) Knowledge-based fault analysis as a central component of quality assurance, *Methods of Operations Research*, vol. 63, pp. 545-554.

Hassoun, M.H. (1995) *Fundamentals of Artificial Neural Networks*, MIT Press, Cambridge, M.A.

Havlicsek, B.L. (1989) Integrating diagnostic knowledge, *IEEE Aerospace and Electronic Systems Magazine*, November, pp. 54-59.

Henry, T.A. (1988) The economics of condition based maintenance, *Proc. of the IMechE Seminar on Condition Based Maintenance of Engines*, 1 March 1988, London, UK.

Hill, J. and Smith, R. (1988), Integrating machinery condition monitoring into maintenance management, *Noise and Vibration Control Worldwide*, 19(8), 248-251.

Hill, J. and Banes, N. (1988) Turning around condition monitoring, *Processing*, 34(4), 31-36.

Hoh, S.M. (1992) *Condition Monitoring and Fault Diagnosis for CNC Machine Tools*, PhD thesis, University of Wales, College of Cardiff, UK.

Ishida, Y., Adachi, N. and Tokumaru, H. (1985) Topological approach to failure diagnosis of large scale systems, *IEEE Transactions on Systems, Man, and Cybernetics*, SMC-15(3), 327-333.

Kramer, M.A. and Leonard, J.A. (1990) Diagnosis using backpropagation neural networks - Analysis and criticism, *Computers and Chemical Engineering*, 14(12), 1323-1338.

Lackinger, F. and Nejdl, W. (1993) A model based troubleshooter based on qualitative reasoning, *IEEE Expert*, Feb., pp. 33-40.

Lee, W. (1990) *A Hybrid Approach to Generic Diagnosis Model for Computer Integrated Manufacturing System*, Ph.D thesis, University of Louisville.

Lippmann, R. (1989) Pattern classification using neural networks, *IEEE Communication Magazine*, pp. 47-64.

Loynes, D. (1981) The influence on condition monitoring on downtime, *Terotechnica*, 1(3), 175-183.

Marko, K.A., James, J., Dosdall, J. and Murphy, J. (1989) Automotive control system diagnostics using neural nets for rapid pattern classification of large data sets, *IEEE Int. Joint Conference on Neural Networks*, Washington, June 18-22, vol. 2 pp. 13-16.

Martin, K.F. (1994) A review by discussion of condition monitoring and fault diagnosis in machine tools, *Int. J. of Machine Tools and Manufacture*, 34(4), 527-551.

Martin, K.F. and Marzi, M.H. (1993) Neural network solution to coolant system diagnosis, *Profitable Condition Monitoring*, Rao, B.K.N. (ed.), Kluwer, Boston, MA, pp. 217-227.

Marzi M.H. and Martin, K.F. (1991) Artificial neural networks in condition monitoring and fault diagnosis, *Proc. of the Int. AMSE Conference on Neural Networks, Methodologies and Applications*, San Diego, May 29-31, California, pp. 113-124.

Mauer, G.F. (1990) On-line cylinder fault diagnostics for internal combustion engines, *IEEE Transactions on Industrial Electronics*, 37(3), 221-226.

Milne, R. (1987) Strategies for diagnosis, *IEEE Transactions on Systems, Man, and Cybernetics*, Special issue on causal and diagnostic reasoning, SMC-17(3), 333-339.

Mitchell, J. (1989) Diagnostic maintenance expert system for the hydraulic subsystem of a continues miner, *IEEE Transactions on Industry Applications*, 25(5), September/October 1989, 841-845.

Monostori, L. (1993) A step towards intelligent manufacturing: modelling and monitoring of manufacturing processes through artificial neural networks, *Annals of the CIRP*, 42(1), 485-488.

Nicholls, C. (1986) Condition monitoring using data collectors, *Chartered Mechanical Engineer*, 33(5), May, 50-51.

Odryna, P. and Strojwas, A.J. (1985) PROD: A VLSI fault diagnostic system, *IEEE Design and Test*, December, pp. 27-35.

Peng, Y. and Reggia, J.A. (1987) A probabilistic casual model for diagnostic problem solving- Part II, *IEEE Transactions on Systems, Man, and Cybernetics*, SMC-17(3), 395-406.

Pflueger, K.W. (1989) Hybrid diagnostic strategy for an expert system controlled automatic test system (EXATS), *IEEE Aerospace and Electronic Systems Magazine*, October, pp. 25-30.

Ronchi, J., Butera, G. and Giudice, D.L. (1990) KANT: An expert system for telediagnosis, *Practical Experience in Building Expert Systems*, Branur, M. (ed.), Wiley, Chichester, UK, pp. 147-162.

Sobajic, D.J., Pao, Y.H. and Dolce, J. (1989) On-line monitoring and diagnosis of power system operating conditions using artificial neural networks, *ISCAS 89,*

22nd IEEE Int. Symposium on Circuits and Systems, Portland, OR, Vol. 3, pp. 2243-2246.

Spur, G. and Specht, D. (1992) Knowledge engineering in manufacturing, *Robotics and Computer Integrated Manufacturing*, 9(4/5), 303-309.

Stells, L. (1989) Diagnosis with a function-fault model, *Applied Artificial Intelligence*, 3(2-3), pp. 213-237.

Szafarczyk, M. (ed.) (1994) *Automatic Supervision in Manufacturing*, Springer-Verlag, London.

Venkatasubramanian, V. and Chan, K. (1989) Neural network methodology for process fault diagnosis, *AIChE Journal*, 35(12), 1993-2002.

VerDuin, W.H. (1990) Neural networks for diagnosis and control, *Journal of Neural Computing*, 1(3), 46-52.

Watanabe, K., Matsuura, I., Abe, M., Kubota, M. and Himmelblau, D. (1989) Incipient fault diagnosis of chemical processes via artificial neural networks, *AIChE Journal*, 35(11), 1803-1812.

Watton, J. (1992) *Condition Monitoring and Fault Diagnosis in Fluid Power Systems*, Ellis Horwood, Chichester, UK.

Wasserman, P.D., Unal, A. and Haddad, S. (1991) Neural network on-line machine condition monitoring systems, *Proc. of the Artificial Neural Networks in Engineering, ANNIE'91*, 10-13 November, St. Louis, Missouri, USA, pp. 693-699.

West, D.A.L. (1988) Condition monitoring in manufacturing - History and overview, *Proc. of the Institution of Production Engineers Seminar on Condition Monitoring in Manufacturing*, 16 June, UWIST, Cardiff, UK

Wilkinson, A.J. (1985) Mind: An inside look at an expert system for electronic diagnosis, *IEEE Design and Test*, August, pp. 69-77.

Williams, J.H., Davies, A. and Drake, P.R. (1994) *Condition-based Maintenance and Machine Diagnostics*, Chapman and Hall, London.

Yang, J.O. and Chang, S.H. (1989) A diagnostic expert system for a nuclear power plant based on the hybrid knowledge approach, *IEEE Transactions on Nuclear Science*, 36(6), 2450-2458.

Yudkin, R.O. (1988) On testing communication networks, *Journal on Selected Areas in Communications*, 6(5), 805-812.

Author Index

Subject Index